NONCOMMUTATIVE DISTRIBUTIONS

MONOGRAPHS AND TEXTBOOKS IN PURE AND APPLIED MATHEMATICS

1. *K. Yano,* Integral Formulas in Riemannian Geometry (1970)
2. *S. Kobayashi,* Hyperbolic Manifolds and Holomorphic Mappings (1970)
3. *V. S. Vladimirov,* Equations of Mathematical Physics (A. Jeffrey, ed.; A. Littlewood, trans.) (1970)
4. *B. N. Pshenichnyi,* Necessary Conditions for an Extremum (L. Neustadt, translation ed.; K. Makowski, trans.) (1971)
5. *L. Narici et al.,* Functional Analysis and Valuation Theory (1971)
6. *S. S. Passman,* Infinite Group Rings (1971)
7. *L. Dornhoff,* Group Representation Theory. Part A: Ordinary Representation Theory. Part B: Modular Representation Theory (1971, 1972)
8. *W. Boothby and G. L. Weiss, eds.,* Symmetric Spaces (1972)
9. *Y. Matsushima,* Differentiable Manifolds (E. T. Kobayashi, trans.) (1972)
10. *L. E. Ward, Jr.,* Topology (1972)
11. *A. Babakhanian,* Cohomological Methods in Group Theory (1972)
12. *R. Gilmer,* Multiplicative Ideal Theory (1972)
13. *J. Yeh,* Stochastic Processes and the Wiener Integral (1973)
14. *J. Barros-Neto,* Introduction to the Theory of Distributions (1973)
15. *R. Larsen,* Functional Analysis (1973)
16. *K. Yano and S. Ishihara,* Tangent and Cotangent Bundles (1973)
17. *C. Procesi,* Rings with Polynomial Identities (1973)
18. *R. Hermann,* Geometry, Physics, and Systems (1973)
19. *N. R. Wallach,* Harmonic Analysis on Homogeneous Spaces (1973)
20. *J. Dieudonné,* Introduction to the Theory of Formal Groups (1973)
21. *I. Vaisman,* Cohomology and Differential Forms (1973)
22. *B.-Y. Chen,* Geometry of Submanifolds (1973)
23. *M. Marcus,* Finite Dimensional Multilinear Algebra (in two parts) (1973, 1975)
24. *R. Larsen,* Banach Algebras (1973)
25. *R. O. Kujala and A. L. Vitter, eds.,* Value Distribution Theory: Part A; Part B: Deficit and Bezout Estimates by Wilhelm Stoll (1973)
26. *K. B. Stolarsky,* Algebraic Numbers and Diophantine Approximation (1974)
27. *A. R. Magid,* The Separable Galois Theory of Commutative Rings (1974)
28. *B. R. McDonald,* Finite Rings with Identity (1974)
29. *J. Satake,* Linear Algebra (S. Koh et al., trans.) (1975)
30. *J. S. Golan,* Localization of Noncommutative Rings (1975)
31. *G. Klambauer,* Mathematical Analysis (1975)
32. *M. K. Agoston,* Algebraic Topology (1976)
33. *K. R. Goodearl,* Ring Theory (1976)
34. *L. E. Mansfield,* Linear Algebra with Geometric Applications (1976)
35. *N. J. Pullman,* Matrix Theory and Its Applications (1976)
36. *B. R. McDonald,* Geometric Algebra Over Local Rings (1976)
37. *C. W. Groetsch,* Generalized Inverses of Linear Operators (1977)
38. *J. E. Kuczkowski and J. L. Gersting,* Abstract Algebra (1977)
39. *C. O. Christenson and W. L. Voxman,* Aspects of Topology (1977)
40. *M. Nagata,* Field Theory (1977)
41. *R. L. Long,* Algebraic Number Theory (1977)
42. *W. F. Pfeffer,* Integrals and Measures (1977)
43. *R. L. Wheeden and A. Zygmund,* Measure and Integral (1977)
44. *J. H. Curtiss,* Introduction to Functions of a Complex Variable (1978)
45. *K. Hrbacek and T. Jech,* Introduction to Set Theory (1978)
46. *W. S. Massey,* Homology and Cohomology Theory (1978)
47. *M. Marcus,* Introduction to Modern Algebra (1978)
48. *E. C. Young,* Vector and Tensor Analysis (1978)
49. *S. B. Nadler, Jr.,* Hyperspaces of Sets (1978)
50. *S. K. Segal,* Topics in Group Kings (1978)
51. *A. C. M. van Rooij,* Non-Archimedean Functional Analysis (1978)
52. *L. Corwin and R. Szczarba,* Calculus in Vector Spaces (1979)

53. *C. Sadosky,* Interpolation of Operators and Singular Integrals (1979)
54. *J. Cronin,* Differential Equations (1980)
55. *C. W. Groetsch,* Elements of Applicable Functional Analysis (1980)
56. *I. Vaisman,* Foundations of Three-Dimensional Euclidean Geometry (1980)
57. *H. I. Freedan,* Deterministic Mathematical Models in Population Ecology (1980)
58. *S. B. Chae,* Lebesgue Integration (1980)
59. *C. S. Rees et al.,* Theory and Applications of Fourier Analysis (1981)
60. *L. Nachbin,* Introduction to Functional Analysis (R. M. Aron, trans.) (1981)
61. *G. Orzech and M. Orzech,* Plane Algebraic Curves (1981)
62. *R. Johnsonbaugh and W. E. Pfaffenberger,* Foundations of Mathematical Analysis (1981)
63. *W. L. Voxman and R. H. Goetschel,* Advanced Calculus (1981)
64. *L. J. Corwin and R. H. Szcarba,* Multivariable Calculus (1982)
65. *V. I. Istrăţescu,* Introduction to Linear Operator Theory (1981)
66. *R. D. Järvinen,* Finite and Infinite Dimensional Linear Spaces (1981)
67. *J. K. Beem and P. E. Ehrlich,* Global Lorentzian Geometry (1981)
68. *D. L. Armacost,* The Structure of Locally Compact Abelian Groups (1981)
69. *J. W. Brewer and M. K. Smith, eds.,* Emily Noether: A Tribute (1981)
70. *K. H. Kim,* Boolean Matrix Theory and Applications (1982)
71. *T. W. Wieting,* The Mathematical Theory of Chromatic Plane Ornaments (1982)
72. *D. B.Gauld,* Differential Topology (1982)
73. *R. L. Faber,* Foundations of Euclidean and Non-Euclidean Geometry (1983)
74. *M. Carmeli,* Statistical Theory and Random Matrices (1983)
75. *J. H. Carruth et al.,* The Theory of Topological Semigroups (1983)
76. *R. L. Faber,* Differential Geometry and Relativity Theory (1983)
77. *S. Barnett,* Polynomials and Linear Control Systems (1983)
78. *G. Karpilovsky,* Commutative Group Algebras (1983)
79. *F. Van Oystaeyen and A. Verschoren,* Relative Invariants of Rings (1983)
80. *I. Vaisman,* A First Course in Differential Geometry (1984)
81. *G. W. Swan,* Applications of Optimal Control Theory in Biomedicine (1984)
82. *T. Petrie and J. D. Randall,* Transformation Groups on Manifolds (1984)
83. *K. Goebel and S. Reich,* Uniform Convexity, Hyperbolic Geometry, and Nonexpansive Mappings (1984)
84. *T. Albu and C. Năstăsescu,* Relative Finiteness in Module Theory (1984)
85. *K. Hrbacek and T. Jech,* Introduction to Set Theory: Second Edition (1984)
86. *F. Van Oystaeyen and A. Verschoren,* Relative Invariants of Rings (1984)
87. *B. R. McDonald,* Linear Algebra Over Commutative Rings (1984)
88. *M. Namba,* Geometry of Projective Algebraic Curves (1984)
89. *G. F. Webb,* Theory of Nonlinear Age-Dependent Population Dynamics (1985)
90. *M. R. Bremner et al.,* Tables of Dominant Weight Multiplicities for Representations of Simple Lie Algebras (1985)
91. *A. E. Fekete,* Real Linear Algebra (1985)
92. *S. B. Chae,* Holomorphy and Calculus in Normed Spaces (1985)
93. *A. J. Jerri,* Introduction to Integral Equations with Applications (1985)
94. *G. Karpilovsky,* Projective Representations of Finite Groups (1985)
95. *L. Narici and E. Beckenstein,* Topological Vector Spaces (1985)
96. *J. Weeks,* The Shape of Space (1985)
97. *P. R. Gribik and K. O. Kortanek,* Extremal Methods of Operations Research (1985)
98. *J.-A. Chao and W. A. Woyczynski, eds.,* Probability Theory and Harmonic Analysis (1986)
99. *G. D. Crown et al.,* Abstract Algebra (1986)
100. *J. H. Carruth et al.,* The Theory of Topological Semigroups, Volume 2 (1986)
101. *R. S. Doran and V. A. Belfi,* Characterizations of C*-Algebras (1986)
102. *M. W. Jeter,* Mathematical Programming (1986)
103. *M. Altman,* A Unified Theory of Nonlinear Operator and Evolution Equations with Applications (1986)
104. *A. Verschoren,* Relative Invariants of Sheaves (1987)
105. *R. A. Usmani,* Applied Linear Algebra (1987)
106. *P. Blass and J. Lang,* Zariski Surfaces and Differential Equations in Characteristic p > 0 (1987)
107. *J. A. Reneke et al.,* Structured Hereditary Systems (1987)

108. *H. Busemann and B. B. Phadke,* Spaces with Distinguished Geodesics (1987)
109. *R. Harte,* Invertibility and Singularity for Bounded Linear Operators (1988)
110. *G. S. Ladde et al.,* Oscillation Theory of Differential Equations with Deviating Arguments (1987)
111. *L. Dudkin et al.,* Iterative Aggregation Theory (1987)
112. T. Okubo, *Differential Geometry* (1987)
113. *D. L. Stancl and M. L. Stancl,* Real Analysis with Point-Set Topology (1987)
114. *T. C. Gard,* Introduction to Stochastic Differential Equations (1988)
115. *S. S. Abhyankar,* Enumerative Combinatorics of Young Tableaux (1988)
116. *H. Strade and R. Farnsteiner,* Modular Lie Algebras and Their Representations (1988)
117. *J. A. Huckaba,* Commutative Rings with Zero Divisors (1988)
118. *W. D. Wallis,* Combinatorial Designs (1988)
119. *W. Więsław,* Topological Fields (1988)
120. *G. Karpilovsky,* Field Theory (1988)
121. *S. Caenepeel and F. Van Oystaeyen,* Brauer Groups and the Cohomology of Graded Rings (1989)
122. *W. Kozlowski,* Modular Function Spaces (1988)
123. *E. Lowen-Colebunders,* Function Classes of Cauchy Continuous Maps (1989)
124. *M. Pavel,* Fundamentals of Pattern Recognition (1989)
125. *V. Lakshmikantham et al.,* Stability Analysis of Nonlinear Systems (1989)
126. *R. Sivaramakrishnan,* The Classical Theory of Arithmetic Functions (1989)
127. *N. A. Watson,* Parabolic Equations on an Infinite Strip (1989)
128. *K. J. Hastings,* Introduction to the Mathematics of Operations Research (1989)
129. *B. Fine,* Algebraic Theory of the Bianchi Groups (1989)
130. *D. N. Dikranjan et al.,* Topological Groups (1989)
131. *J. C. Morgan II,* Point Set Theory (1990)
132. *P. Biler and A. Witkowski,* Problems in Mathematical Analysis (1990)
133. *H. J. Sussmann,* Nonlinear Controllability and Optimal Control (1990)
134. *J.-P. Florens et al.,* Elements of Bayesian Statistics (1990)
135. *N. Shell,* Topological Fields and Near Valuations (1990)
136. *B. F. Doolin and C. F. Martin,* Introduction to Differential Geometry for Engineers (1990)
137. *S. S. Holland, Jr.,* Applied Analysis by the Hilbert Space Method (1990)
138. *J. Okniński,* Semigroup Algebras (1990)
139. *K. Zhu,* Operator Theory in Function Spaces (1990)
140. *G. B. Price,* An Introduction to Multicomplex Spaces and Functions (1991)
141. *R. B. Darst,* Introduction to Linear Programming (1991)
142. *P. L. Sachdev,* Nonlinear Ordinary Differential Equations and Their Applications (1991)
143. *T. Husain,* Orthogonal Schauder Bases (1991)
144. *J. Foran,* Fundamentals of Real Analysis (1991)
145. *W. C. Brown,* Matrices and Vector Spaces (1991)
146. *M. M. Rao and Z. D. Ren,* Theory of Orlicz Spaces (1991)
147. *J. S. Golan and T. Head,* Modules and the Structures of Rings (1991)
148. *C. Small,* Arithmetic of Finite Fields (1991)
149. *K. Yang,* Complex Algebraic Geometry (1991)
150. *D. G. Hoffman et al.,* Coding Theory (1991)
151. *M. O. González,* Classical Complex Analysis (1992)
152. *M. O. González,* Complex Analysis (1992)
153. *L. W. Baggett,* Functional Analysis (1992)
154. *M. Sniedovich,* Dynamic Programming (1992)
155. *R. P. Agarwal,* Difference Equations and Inequalities (1992)
156. *C. Brezinski,* Biorthogonality and Its Applications to Numerical Analysis (1992)
157. *C. Swartz,* An Introduction to Functional Analysis (1992)
158. *S. B. Nadler, Jr.,* Continuum Theory (1992)
159. *M. A. Al-Gwaiz,* Theory of Distributions (1992)
160. *E. Perry,* Geometry: Axiomatic Developments with Problem Solving (1992)
161. *E. Castillo and M. R. Ruiz-Cobo,* Functional Equations and Modelling in Science and Engineering (1992)
162. *A. J. Jerri,* Integral and Discrete Transforms with Applications and Error Analysis (1992)
163. *A. Charlier et al.,* Tensors and the Clifford Algebra (1992)

164. *P. Biler and T. Nadzieja*, Problems and Examples in Differential Equations (1992)
165. *E. Hansen*, Global Optimization Using Interval Analysis (1992)
166. *S. Guerre-Delabrière*, Classical Sequences in Banach Spaces (1992)
167. *Y. C. Wong*, Introductory Theory of Topological Vector Spaces (1992)
168. *S. H. Kulkarni and B. V. Limaye*, Real Function Algebras (1992)
169. *W. C. Brown*, Matrices Over Commutative Rings (1993)
170. *J. Loustau and M. Dillon*, Linear Geometry with Computer Graphics (1993)
171. *W. V. Petryshyn*, Approximation-Solvability of Nonlinear Functional and Differential Equations (1993)
172. *E. C. Young*, Vector and Tensor Analysis: Second Edition (1993)
173. *T. A. Bick*, Elementary Boundary Value Problems (1993)
174. *M. Pavel*, Fundamentals of Pattern Recognition: Second Edition (1993)
175. *S. A. Albeverio et al.*, Noncommutative Distributions (1993)
176. *W. Fulks*, Complex Variables (1993)
177. *M. M. Rao*, Conditional Measures and Applications (1993)

Additional Volumes in Preparation

NONCOMMUTATIVE DISTRIBUTIONS

Unitary Representation of Gauge Groups and Algebras

Sergio A. Albeverio
Ruhr-Universität Bochum
Bochum, Germany

Raphael J. Høegh-Krohn
University of Oslo
Oslo, Norway

Jean A. Marion
Daniel H. Testard
Bruno S. Torrésani
Université Aix-Marseille II and
Centre de Physique Théorique (CNRS)
Marseille, France

Marcel Dekker, Inc. New York • Basel • Hong Kong

Library of Congress Cataloging-in-Publication Data

Noncommutative distributions : unitary representation of gauge groups and algebras / by S. Albeverio ... [et al.].
p. cm. -- (Monographs and textbooks in pure and applied mathematics ; 175)
Includes bibliographical references.
ISBN 0-8247-9131-2 (acid-free)
1. Lie algebras. 2. Quantum field theory. 3. Representations of groups. 4. Algebra of currents. 5. Mathematical physics.
I. Albeverio, Sergio. II. Series.
QC20.7.L54N65 1993
512'.55--dc20 93-9942
CIP

The publisher offers discounts on this book when ordered in bulk quantities. For more information, write to Special Sales/Professional Marketing at the address below.

This book is printed on acid-free paper.

Marcel Dekker, Inc.
270 Madison Avenue, New York, New York 10016

Current printing (last digit):
10 9 8 7 6 5 4 3 2 1

PRINTED IN THE UNITED STATES OF AMERICA

Preface

It is extremely sad that Raphael Høegh-Krohn's too short life (February 10, 1938 to January 24, 1988) did not permit him to witness the publication of this book. A great part of the book was practically finished when Raphael died. The shock of his departure made us hesitate for a long time before publishing the book, and we missed him terribly when we decided to proceed. Among Raphael's many projects, the investigation of representations of infinite dimensional Lie groups occupied a privileged position. Since his initial work on the topic with one of us (S. A., 1975), where we first discovered—at about the same time and independently of Gelfand et al. and Ismagilov—the energy representation of groups of mapping from a manifold into a compact Lie group, he kept thinking about questions concerning the classification of irreducible representations of infinite dimensional Lie groups. His thinking was greatly stimulated by his meeting with Anatoli Vershik in June 1976 at a conference on information theory organized by R. Dobrushin and later on by his meeting with Israel Gelfand and M. Graev in 1979 where Raphael reported on his work on the topics at Gelfand's seminar.

These contacts also stimulated our combined research, which grew out of a joint seminar on the topics during S. Albeverio's and R. Høegh-Krohn's stay at the Centre de Physique Théorique, CNRS, Marseille

(1977–78). This collaboration continued during subsequent stays in Marseille, where Raphael was associated for many years with the Université de Provence and the Centre de Physique Théorique. Raphael also lectured on the topics in Marseille and Oslo and directed the Ph.D. studies of one of us (B. T.). He also discussed these topics with other students in Oslo, in particular Steinar Johannesen, Helge Holden, and Terje Wahl. We developed the idea of the present book in Marseille and during a common stay at the Center for Interdisciplinary Research (ZiF) and the University of Bielefeld within the framework of a Research Year in Mathematics and Physics (1983–84). Work on the book and connected research was continued during other common stays at various places. Besides those stated, we would like to mention the mathematics departments of Oslo University and the Universities of Bielefeld and Bochum.

We are most grateful to all these institutions, and particularly to our friends at those institutions, Jean Bellissard, Philippe Blanchard, Philippe Combe, Daniel Kastler, Mohammed Mebkhout, Ludwig Streit, Roger Rodriguez, Madeleine Sirugue-Collin, Raymond Stora, and the late Michael Sirugue for their generous hospitality and stimulating discussions.

Lectures on the topics of the book were given on various occasions, in particular at the IAMP Meeting in Rome (1977) (organized by G. F. Dell' Antonio, S. Doplicher, and G. Jona-Lasinio), a Bielefeld Workshop in the Framework of the Mathematization Project (1980) (organized by S. Albeverio, Ph. Blanchard, and L. Streit), Kyoto International Conference on Markov Processes and Analysis (1981) (organized by M. Fukushima and H. Kunita), the Ascona/Locarno Meetings on Stochastic Processes in Classical and Quantum Systems (1985) and Stochastic Processes, Physics and Geometry I, II (1989, 1991) (organized by S. Albeverio, G. Casati, U. Cattaneo, D. Merlini, and R. Moresi), the ZiF Conferences Stochastic Processes in Mathematics and Physics II and Trends and Developments in the Eighties (1985) (organized by S. Albeverio, Ph. Blanchard, and L. Streit), the Oslo R. Høegh-Krohn Memorial Conference (organized by S. Albeverio, J. E. Fenstad, H. Holden, and T. Lindstrøm), the Conference on Operator Algebras and Group Representations (organized in September 1980 by D. Voiculescu) at Neptun (Romania), in the form of a course given at the Université d'Abidjan (1989), and a lecture at the Conference on Probability Measures on Groups organized by H. Heyer in Oberwolfach (1990).

We are also grateful to a number of other persons either for interesting discussions or for stimulating our work in one way or the other. In particular, we would like to thank Anatoli Vershik, Israel Gelfand, M.

Graev, Nolan Wallach, Ronal Dobrushin, Robert Minlos, Yasha Sinai, Victor Kac, Zbigniev Haba, Witold Karwowski, Masatoshi Fukushima, Herbert Heyer, G. O. S. Ekhaguere, Gian Fausto Dell' Antonio. The financial support of the Volkswagenstiftung, Alexander von Humboldt-Stiftung, DFG, DAAD, CNRS, NAVF, Ministère de la Recherche et de la Technologie at various stages is gratefully acknowledged.

Last but not least, we would like to thank Regine Kirchhoff, Suzanne Léon, Brigitte Richter, Antoinette Sueur, Ursula Weber, and the publisher's staff for their skillful help.

Sergio A. Albeverio
Jean A. Marion
Daniel H. Testard
Bruno S. Torrésani

Contents

NONCOMMUTATIVE DISTRIBUTIONS

Introduction

This book is concerned with noncommutative distribution theory, more particularly with the representation theory of certain infinite-dimensional Lie groups called gauge groups and associated Lie algebras. Let us illustrate a bit the concepts involved and motivate the interest in their study.

Noncommutative distribution theory should be roughly speaking distribution theory (i.e., the study of generalized functions [GeSh, GeVi, GeGrVi, GeGrPS]) extended to the noncommutative case, for instance in the following sense: to each real-valued distribution T on a manifold X we can associate the $U(1) \cong S^1$-valued map $e^{i\langle T,\cdot\rangle}$ from a test function space $\mathcal{T}(X)$, so that for any $\varphi \in \mathcal{T}(X)$, $e^{i\langle T,\varphi\rangle}$ belongs to the unit complex circumference $\{z \in \mathbb{C} | \, |z| = 1\}$. $e^{i\langle T,\cdot\rangle}$ can be looked upon as a character of the infinite-dimensional group G^X (pointwise multiplication) of maps from $\mathcal{T}(X)$ (assumed to be an algebra) into $G \equiv \mathbb{R}$ (in the obvious sense that for each fixed $\varphi \in \mathcal{T}(\mathbb{R})$, $e^{i\langle T,\alpha\varphi\rangle}$, $\alpha \in \mathbb{R}$, is a character of the abelian group $\mathbb{R}$).

Vice versa, the knowledge of the character $e^{i\langle T,\varphi\rangle}$ for all $\varphi \in \mathcal{T}(X)$ characterizes the distribution T. But characters are in one-to-one correspondence with irreducible continuous unitary representations of abelian groups, hence to find a possible noncommutative extension of the concept of distribution one can look into (equivalence classes of) unitary representations of (nonabelian) groups of (smooth) mappings $\mathcal{D}(X, G)$, where G is a (Lie) group ($\mathfrak{D}(X, G)$ being given the group structure obtained by taking pointwise multiplication). This is a point of view taken originally by Gelfand and his collaborators, who introduced the subject into the mathematical literature (see, e.g., [GeGr]).

From this point of view, noncommutative distribution theory is intimately connected with the study of infinite-dimensional Lie groups $\mathfrak{D}(X, G)$ of mappings from a manifold X into a Lie group G, and their representations. It turns out that the study of such groups and the associated (infinite-dimensional) Lie algebras is a topic of great interest in itself and because of its numerous and fruitful connections with many domains of mathematics and its applications (especially in theoretical physics).

In fact, the creator of the theory of Lie groups, the Norwegian Sophus Lie, already looked upon Lie groups as general groups of symmetries of geometric objects or analytic objects and by this he was from the beginning interested in infinite-dimensional as well as finite-dimensional Lie groups.

Lie himself introduced the so-called infinitesimal methods in the study of Lie groups, namely the study of the corresponding infinite- (resp. finite-) dimensional Lie algebras. One passes from Lie algebras to Lie groups by the exponential mapping, whose properties are well known, at least in the finite-dimensional case (see, e.g., [FrdV, He, Va] and, for certain infinite-dimensional cases, [Om, Mic, CiM]).

For this reason we switch often in our further discussion between Lie groups and Lie algebras, since, at least in principle, results about the one structure can be looked upon as results about the other and vice versa.

Simple finite-dimensional Lie algebras over the field $\mathbb{C}$ of complex numbers were classified already at the end of last century, mainly due to work by Killing and Cartan [Kil, Car]. As well known, this is at the basis of the study of general finite-dimensional Lie groups, for which there are structure theorems (roughly speaking one reduces the study of the groups and their representations to the one of simple compact Lie groups, abelian Lie groups and nilpotent Lie groups) (see, e.g., [Che, Dix, Hari, Dyn, He, Bou]). In contrast to the case of finite-dimensional Lie groups and algebras, whose theory had a steady development since Lie, Killing and Cartan's pioneering work, the theory of infinite-dimensional Lie groups and algebras did not develop much until the mid-1960s, despite the fact that Cartan himself had extended already at the beginning of this century his classification of finite-dimensional simple Lie algebras to the case of simple infinite-dimensional Lie algebras of vector fields over a finite-dimensional space [Car].

These algebras correspond to groups of diffeomorphisms of a manifold. The study has been in fact extended only much later (in the mid-1960s) by Gelfand-Fuchs, Shafarevitch, Kirillov and others into the study of infinite-dimensional Lie algebras of vector fields on a finite-dimensional manifold and groups of automorphisms of a manifold (or, more generally, an algebraic variety).

Special attention has been devoted to the study of the group of diffeomorphisms of a manifold and of its representations [Is2, Sh]. This group is also of importance in relation with certain partial differential equations and as such its study has had applications in hydrodynamics and magnetohydrodynamics (e.g., [Arn, MaEbFi, MaW]) and in general relativity, (e.g., [GoMS1, 2, 3, GoSh1, 2, GiWy]).

Other infinite-dimensional Lie groups, e.g., the so-called Geroch group [GeWu] and their representations arise as group of transformations in general relativity (see e.g. [Ju1] and references therein).

A second class of infinite-dimensional Lie groups (resp. Lie algebras) which is presently rather well studied is precisely the class $\mathfrak{D}(X, G)$ (resp. $\mathfrak{D}(X, \mathcal{G})$) of mappings of a given manifold X into a Lie group G (resp. Lie algebra $\mathcal{G}$).

This is the class to whose study we shall dedicate this book. Before we come to discuss this class let us mention however the study of some other types of infinite-dimensional Lie groups and algebras.

A natural class to be studied is the infinite-dimensional analog of classical finite-dimensional matrix Lie groups and algebras where the matrices are replaced by operators in infinite-dimensional Hilbert (or Banach) spaces. The cases $U(\infty)$, $SO(\infty)$, $Sp(\infty)$ have been studied, e.g., in Kirillov [Ki1–4], who classified all their irreducible unitary representations (see also, e.g., [Ba, Gol, Hi, Ner, Rag, Seg, Sha, Sto, StrV, Pi, OkS, Sm]). See also [KaP1] for a description of a wedge representation of $GL(\infty)$ and applications to vertex operators and Bose-Fermi correspondence.

A construction of representations of Kac-Moody algebras and superalgebras starting from infinite-dimensional CCR and CAR has been given in [FFr].

Unitary representations of $U(p, \infty)$, $SO_0(p, \infty)$, $Sp(p, \infty)$ were studied by Ol'shanskii [Ol].

The study of infinite-dimensional Weyl-Heisenberg groups, equivalent with the study of canonical commutation relations of quantum field theory (an extension of the Weyl-von Neumann study of finite-dimensional commutation relations, see, e.g., [Pu]), has been extensively pursued.

Part of this study is concerned with quantum field theory, e.g., [AHK6, Xi, Jo, WiGa, GeW, Ar1] or string theory [GSW], where the infinite-dimensional Heisenberg algebra is the starting point of the construction of the so-called basic representation of simply laced affine Kac-Moody algebras, via the vertex operator construction [FrK, FrLM, SoT]. (Note that an infinite-dimensional version of the Stone-von Neumann theorem has been obtained in this context [Ka1].)

Another part of the study looks at infinite-dimensional Weyl groups as mathematical object of interest in themselves, namely as realizations of an infinite-dimensional symplectic group, hence an infinite-dimensional version of the Segal-Shale-Weil theory of metaplectic representations (see, e.g., [dlH, Or]). For a general theory of infinite-dimensional Lie groups containing the quoted classes, see also e.g. [Om, Mic, Mil]. Some of these connections are actually related to other infinite-dimensional Lie groups associated with Kac-Moody (affine) Lie algebras, which we discuss extensively.

Let us now comment on the groups, algebras, and representations we are basically concerned with in this book. The groups are those we generically indicated above by $\mathfrak{D}(X, G)$. Such groups are called sometimes—and we will follow this custom—*gauge groups*. The name comes from the fact that such gauge groups arise in the representation theory of gauge quantum fields (as we shall discuss further), in which case X has the interpretation of physical space (or space-time) (often compactified in a suitable way and mostly made "locally Euclidean" by replacing the local Minkowski metric by a corresponding Euclidean one). G has the interpretation of the structure Lie group for the principal fiber bundle whose connection defines the classical gauge field. Quantization of gauge fields corresponds then, in this interpretation, to finding unitary representations of the gauge groups $\mathfrak{D}(X, G)$ (in some separable Hilbert space) which is the state space of the corresponding gauge field theory.

We also note that sometimes the gauge groups are also called *current groups*. This is related to the fact that the corresponding Lie algebras occur in physics (in connection with conservation laws) (especially high-energy physics) under the name *current algebras*; see, e.g., [Tr, MiRa, Ree, Ar2].

Despite the fact that these groups $\mathfrak{D}(X, G)$ are natural and simple extensions of classical finite-dimensional Lie groups and they have the above interesting connections with objects of study in physics, the theory of their representations has been developed only relatively recently and one has still only partial results. It is one of the aims of the present book to give a unified presentation of what is presently known and provide the basis for future work.

Let us discuss a little the particularities of the above gauge groups as infinite-dimensional Lie groups. They do possess a natural Lie algebra

$\mathcal{D}(X, \mathcal{G})$ (the Lie algebra of smooth mappings from X into the Lie algebra $\mathcal{G}$ of G) and an exponential operation (given by pointwise exponentiation), which is locally one-to-one. One of the main problems of the theory of gauge groups consists in constructing "sufficiently nontrivial representations." Let us explain what we mean by "sufficiently nontrivial representations." One can always construct evaluation representations which are compositions of an evaluation of a mapping at points of X and an ordinary representation of G. These evaluation representations are clearly not interesting because they give no information about the differential structure of X or G. From a physical point of view, the representations of gauge groups of interest have to depend not only on evaluations but also on successive derivatives of the mappings. The problem of constructing irreducible representations of gauge groups depending on derivatives of maps corresponds in the commutative case to the one of passing from measures to general distributions on a manifold. Hence in this sense the theory of representations of gauge groups appears as a genuine noncommutative version of the theory of distributions, which was our starting point above.

This book is devoted to the study of examples of constructions extending in the above sense commutative distributions to noncommutative ones. The examples are not deeply connected to the global structure of X and, in reality, we only deal with the local noncommutative distribution problem. An extension of the theory to a global setting would certainly be worthwhile, also in view of connections with "topological quantum field theories." At this point, it is interesting to note that in contrast with the ordinary theory of distributions, in case of measures, the classical extension of a measure from the continuous functions in $C_0(X, G)$ (C_0 standing for continuous functions of compact support) to the Borel functions $\mathcal{B}(X, G)$ may introduce problems. An easy example is provided by the case where $G = S^1$. In this case the groups $C_0(X, S^1)$ and $\mathcal{B}(X, S^1)$ are abelian, and irreducible representations are one-dimensional. Restricting our attention to the unit component in the group $C_0(X, S^1)$, i.e., to the functions of the form $x \mapsto e^{i\psi(x)}$, with $\psi \in C_0(X, \mathbb{R})$, one easily sees that irreducible representations are given by measures such that any connected component of X has a measure in the set $2\pi\mathbb{Z}$. These representations cannot be extended to $\mathcal{B}(X, S^1)$, except in the case where the support of the measure is totally disconnected, i.e., in case of Dirac measures with coefficients in $2\pi\mathbb{Z}$. Nevertheless, if one wants to have a differentiability condition in order to be able to get an infinitesimal representation of the Lie-algebra-valued maps $C_0(X, \mathcal{G})$, then any representation of $C_0(X, G)$ extends to a representation of $\mathcal{B}(X, G)$.

In order to achieve the above program of constructing noncommutative distributions a natural idea is to first consider what is called the G-valued jet bundle associated to the manifold X [GeGrVe] and its representations "of order k" (for the definition see Chapter 1). In this way, the first problem to solve is to find representations of a space of continuous sections of the kth jet bundles which have locally the same structure as $\mathcal{D}(X, G)$ but with G replaced by an extension H of G by a nilpotent group. For instance for $k = 1$, $J^1(X, G)$ is an extension of $\mathcal{D}(X, G)$ by the set $\Omega^1(X, \mathcal{G})$ of $\mathcal{G}$-valued 1-forms on X considered as an abelian group, $\mathcal{D}(X, G)$ acting on $\Omega^1(X, \mathcal{G})$ via the pointwise adjoint representation. As we will see, following work by Gelfand et al. [VeGeGr1, 2, VeK], Guichardet [Gui], Delorme [De1], and others (see also e.g., [Kaz, Wan]), the nontriviality of the 1-cohomology of $J^1(X, G)$ plays an important role in connection with the problem of constructing noncommutative distributions of order 1. In this context it is important to note that the mapping

$$\beta: \mathcal{D}(X, G) \to \Omega^1(X, \mathcal{G}), \qquad \beta(\varphi) = d\varphi\, \varphi^{-1}$$

is a nontrivial 1-cocycle for the pointwise adjoint representation.

The problem of constructing representations of $C(X, H)$, the continuous functions from X to H, where H is the considered extension of G, appears as a suitable generalization of problems of ordinary measure theory. Under conditions making the representation sufficiently regular in order to be able to define an infinitesimal representation of the Lie algebra $\mathcal{D}(X, g)$, a representation of $C(X, H)$ which is continuous with respect to the uniform topology can be extended to the group of Borel functions from X to H. Furthermore it is possible to develop a concept of discrete and continuous parts of the support of a representation (see [Ar2]). In the case of a finite discrete support, one also easily sees that the only possibility for irreducible representations is a finite tensor product of evaluation representations at a finite number of points. This agrees with the classical local structure of ordinary distributions. The case of a continuous support is much more complicated, and we do not have a structure theorem as in the case of a discrete support. The only way to get a construction which seems to agree with the (multiplicative version) of locality is to use the process introduced in [GeGr, PaS1, 2, Str1–3, Math, ArW, Gui], exploit the structure of motion group which was stressed before for $J^1(X, G)$, and use the representations of type S in the terminology of [Gui] and [De1]. All these ideas will be developed in Chapter 2.

Chapters 3 and 4 are devoted to the study of the energy representation of gauge groups $\mathcal{D}(X, G)$. This representation with G compact semisimple, has been found by Ismagilov [Is1, 3], Gelfand, Graev, and Vershik [GeGrVe, VeGeGr3], and Albeverio and Høegh-Krohn [AHK1] (see also [Wa, GaTr]). It is a representation whose cyclic component of its vacuum can be characterized by a spherical function whose Fourier transform is given in terms of the energy Dirichlet form naturally associated with the Killing form on the Lie algebra, by $\exp(-\frac{1}{2}\int_X |d\varphi\ \varphi^{-1}|^2\, dx)$, where $\varphi \in \mathcal{D}(X, G)$, $d\varphi\ \varphi^{-1}$ is the Maurer-Cartan cocycle, $|\ |$ being given by the Euclidean metric on X and the Killing metric on the Lie algebra g of G. The connection with other representations and objects is best understood in the case where X is one-dimensional ($\mathbb{R}_+$, $\mathbb{R}$, or S^1, say). In this case, as shown originally in [AHK1], the representation can be identified with the regular representation given by left or right translation by elements of $\mathcal{D}(X, G)$ of the paths of Brownian motion (resp., in the case of S^1, of Brownian loops). In the case $X = S^1$ this is then the energy representation of the group of loops with values in G (whose infinitesimal representation is a non-highest-weight representation of the corresponding loop algebra).

Due to its expression for dim $X = 1$, as the representation of the group by left translations, it is easy to show that the energy representation is, in this case, not irreducible [AHK1]. In fact its reduction theory can be worked out [AHKT, AHKTV]. This is discussed in Chapter 4. This study involves stochastic analysis on Lie groups. Only almost sure irreducibility of the reduced representations with respect to a Wiener measure is achieved in this way. We also remark that the corresponding object to the energy representation when G is abelian ($G = \mathbb{R}$ or $\mathbb{R}^s$ or T^s) and $X = \mathbb{R}^d$ is the free massless scalar ($\mathbb{R}$) or vector ($\mathbb{R}^s$) or periodic (T^s) Euclidean quantum field over space-time $\mathbb{R}^d$. Such free fields are well-studied Markov generalized random fields ([Ne, AHKZ, Rö]), the building stones for models of Euclidean and relativistic interacting quantum fields, see, e.g., [GlJa, Si, AHK2, AFHKL]. Such fields appear as coordinate maps in the energy representation which in this commutative case is nothing but the Schrödinger representation of the Euclidean fields [AHK1, 2, 3, 5, 6].

From this point of view one can look upon the energy representation for general abelian G as providing a noncommutative extension of such fields to fields with values in Lie groups. In turn this gives an algebraic realization of the concept of nonlinear σ-model (although the representation, rather than the coordinate functions or the field itself, is given) (for literature on nonlinear σ-models see, e.g., [Ha1, 2, GeW, Sk, Za] and references therein).

In the case dim $X \geq 2$ there is no direct connection with Brownian motion and the energy representation is given "only algebraically" in the $L^2(\mu)$-space with respect to the natural Gaussian measure associated with the scalar product given by the energy form $\int |d\varphi\, \varphi^{-1}|(x)^2\, dx$ by translation by the Maurer-Cartan 1-form. There is a well-known unitary equivalence of $L^2(\mu)$ with the Fock space $\mathcal{F}$ over the Sobolev space $H_1^2(\mathbb{R}^d)$; see, e.g., [Gui].

The energy representation in $\mathcal{F}$ takes the form, on exponential or coherent vectors (see, e.g., [KlS, McK, Pe]) of the "exponentiation" (second quantization) of the left multiplication followed by translation by the Maurer-Cartan 1-form. Thus, in this realization, the energy representation has the form of the representation of a Euclidean group.

The question of the reducibility-irreducibility of the energy representation is highly dependent on the dimension d of the underlying manifold X. The irreducibility is known for $d \geq 3$ [GeGrVe, Is1, AHK T, Wa], the method involving support properties of the Gaussian measures μ coupled with estimates first developed in connection with the construction of quantum field models with exponential interaction (Høegh-Krohn model) [AHK3, 4, AGHK]. The same methods also work for $d = 2$ under an additional hypothesis involving the lengths of roots of the Lie algebra g (corresponding to the fact that the so-called :exp $\alpha\varphi$: interaction of quantum field theory exists [AHK3] and is nontrivial for $\alpha^2 < 4\pi$ only for $d = 2$ [AGHK, AHK4, Ku, Wa, O]).

A recent improvement of proofs by Wallach [Wa] is an exact confirmation of this correspondence including coincidence of constants in the two problems.

The construction and study of the energy representation has been extended in [Ma1–4] and [MaT] to much more general situations, in particular to flag manifolds.

Chapter 5 is closely related to Chapters 3 and 4, inasmuch as it is concerned with an extension of the study of the group of mappings $\mathcal{D}(X, G)$, resp. $\mathcal{D}(X, g)$, to a representation of the universal central extension of it (see, e.g., [Kan, Moo1, 2, Mor1–4, PT, Ti, Verm, PrS, Wi, Ga1, 2]). For $X = S^1$, this is the study of affine (or Euclidean) Kac-Moody Lie groups and Lie algebras. The study of such infinite-dimensional loop groups and algebras has had a very extensive development since its inception in 1968. They are special cases of the general Kac-Moody Lie groups and algebras (technically, affine Kac-Moody Lie algebras which are Kac-Moody Lie algebras of height 1). The literature on these subjects is very rich, fortunately there are also some quite extensive surveys of at least a good part of it; see, e.g., [Fr1–3, BäuK, Bot, GoWa, Dun1, 2, Ka1–5, KaP1, KaW, Mat, PrS, Se1, 2, VZ], and references therein (see also the bibliography in [A, Be]). This theory comprehends structure results as well as explicit realizations and the study of certain types of representations. In our presentation we give a unitary representation obtained as equivariant representation of loop groups (based on [Te]), starting from the energy representation.

In the quoted literature another type of representation is discussed extensively, namely the highest weight or positive energy representation (imitating in a sense the Borel-Weil finite-dimensional theory) (note that "positive energy" here has nothing to do with the energy in the name "energy representation").

In the affine case, a complete classification of all unitary positive energy representations has been obtained in [JKa], starting from a careful analysis of the different ways of defining a total ordering on the root system of the considered affine Lie algebra. Apart from the so-called elementary representations (essentially tensor products of representations of a semisimple Lie algebra), there remain two types of representations: the "integrable" representations, in the cases where G is compact, are usually used in physics, and the "exceptional" representations, for $G = SU(n, 1)$, in which the central extension disappears. The link between the exceptional representation of $SU(1, 1)^{s'}$ and continuous tensor products has been studied in [De2, To1].

The literature on applications of (affine) Kac-Moody Lie groups and algebras is very extensive, see e.g. [Go, GoKO, Her, Man, Oli]. Let us also remark that it contains also applications of the Virasoro algebra (the Lie algebra of the universal central extension of the group of diffeomorphisms of the circle [FeiFu]). Let us mention a few types of domains of applications (this list is far from exhaustive):

a. Supersymmetry and gravitation, Yang-Mills theory, chiral models: see, e.g., [Ju2, GeLiWu, Wu1, 2, GeWu, Do1, 2, KZ, San, SpScTroVP].
b. Exactly solvable models of classical nonlinear partial differential equations in two dimensions (soliton equations, KP and MKP hierarchies: see, e.g., [Sa, DJKM, SeW, KaR]; integrable systems in two dimensions [DJKM, ReSe, SaVe, Sa, deV].
c. Two-dimensional statistical mechanics, critical phenomena: see, e.g., [GoO1–3, CIZ, Ca1, 2]; two-dimensional conformal quantum field theory (here the symmetry under the Virasoro algebra plays an essential role in the classification of models): see, e.g., [Bar, BoBCM,

BouN, Bouw, BPZ, Ca1, 2, CoG, CIZ, FrQS, Fu, Go, KasMaQ, KZ, LM, PT, ScW, Tod, Ver].

d. String theory: see, e.g., [AG, Alv, Fiki, FrK, GoO1–3, Kak, KrN, Ro, GSW, Na, FrGZ, AGO, AHKPS, AJPS].
e. Exactly solvable lattice statistical models: see, e.g., [DJMO, Bax]; for relations of such models with braid groups, operator algebras, theta functions, and modular invariance; see also, e.g., [Sch, Pa, FeFröK1, 2, FoNi].
f. Quantum groups [Dr, Ros, Fa1, 2, Wo, Co1, 2, Kas1, 2, Man].
g. Theta functions theory and combinatorics: [Cart, McD, P, Le1, LeW, KaP1].
h. Study of singularities in algebraic geometry: see, e.g., [Sl, Lu].
i. Study of Boson-Fermion correspondence in quantum fields: [KaP1–3, KaR, Fr2, Mi1].
j. Kac-Moody superalgebras, and related conformal algebras: [FFr, Ka4, Fr3].
k. Super Kac-Moody algebras and super conformal algebras: [BaMPZ, GoKo, KaTo, CFRS, Ad, Tr].
l. Extended conformal algebras: [Zam, BaBSS1, 2, Rav].

In Chapter 5, we briefly sketch the essential features of the theory of affine Kac-Moody algebras, referring to [Ka1] for further developments. The corresponding highest weight representation theory is also presented, in connection with applications to string theory and two-dimensional conformal field theory.

For the case of an X-space of arbitrary dimension, we develop the theory of quasisimple Lie algebras (essentially corresponding to $X = T^\nu$, the ν-dimensional torus) [HKT], in order to be able to construct a highest weight representation theory. It turns out that quasisimple Lie algebras can be classified (generalizing the Kac-Moody classification), and that some of them can be realized as Lie algebras of mappings from a torus into a semisimple Lie algebra (gauge algebra). However, the corresponding unitary highest weight representation theory turns out to be fairly poor in the following sense: the integrable representations of Kac-Moody algebras (corresponding to compact G) do not generalize nontrivially for higher-dimensional X. Nevertheless, only in the case $G = SU(n, 1)$, there exists a nontrivial unitary highest weight representation of $\mathfrak{D}(X, \mathcal{G})$, called the exceptional representation. Moreover, this representation is nothing but the infinitesimal representation of the continuous tensor product representations discussed in Chapter 1 [De2, To1–3].

This clearly indicates that for $\dim(X) > 1$ and $G \neq SU(n, 1)$, it is necessary to consider non-highest-weight representations of $\mathfrak{D}(X, \mathcal{G})$, or nonlinear representations, to get nontrivial unitary representations.

BIBLIOGRAPHY

[A] R. Arcuri, Ph.D. thesis, Imperial College (1990).

[Ad] M. Ademollo et al., *Phys. Lett.*, *62B*:105 (1976).

[AFHKL] S. Albeverio, J. E. Fenstad, R. Høegh-Krohn, and T. Lindstrøm, *Nonstandard Methods in Stochastic Analysis and Mathematical Physics*, Academic Press, New York (1986) (translation into Russian: Mir (1990).

[AG] L. Alvarez-Gaumé, *Operator Methods in String Theory*, in *Nonperturbative Quantum Field Theory*, ('t Hooft et al., eds.) NATO ASI series (1989).

[AGHK] S. Albeverio, G. Gallavotti, and R. Høegh-Krohn, *Comm. Math. Phys.*, *70*:187 (1979).

[AGO] R. Arcuri, J. Gomez, and D. Olive, *Nucl. Phys.*, *B285*:327 (1987).

[AHK1] S. Albeverio and R. Høegh-Krohn, *Comp. Math.*, *36*:37 (1978).

[AHK2] S. Albeverio and R. Høegh-Krohn, *Stochastic Analysis and Applications* (M. Pinsky, ed.), Marcel Dekker, New York, (pp. 1–98) (1984).

[AHK3] S. Albeverio and R. Høegh-Krohn, *J. Funct. Analy.*, *16*:39 (1974).

[AHK4] S. Albeverio and R. Høegh-Krohn, *Quantum Fields—Algebras, Processes* (L. Streit, ed.), Springer-Verlag, Wien, pp. 331–335 (1980).

[AHK5] S. Albeverio and R. Høegh-Krohn, *Mathematical Problems in Theoretical Physics* (G. Dell'Antonio, S. Doplicher, and G. Jona-Lasinio, eds.), Lecture Notes in Physics, vol. 80, Springer-Verlag, New York, pp. 279–302 (1978).

[AHK6] S. Albeverio and R. Høegh-Krohn, Mathematical Methods of Quantum Field Theory, CNRS 248, Paris, pp. 11–59 (1976).

[AHK7] S. Albeverio and R. Høegh-Krohn, *Comm. Math. Phys.*, *68*:95 (1979).

[AHKPS] S. Albeverio, R. Høegh-Krohn, S. Paycha, and S. Scarlatti, *Acta Applic. Math.*, *26*:103 (1992).

[AHKZ] S. Albeverio, R. Høegh-Krohn, and B. Zegarlinski, *Comm. Math. Phys.*, *123*:377 (1989).

[AJPS] S. Albeverio, J. Jost, S. Paycha, and S. Scarlatti, A mathematical introduction to string theory-variational problems, geometric and probabilistic methods, Bochum Preprint (1993).

[AHKT] S. Albeverio, R. Høegh-Krohn, and D. Testard, *J. Funct. Anal.*, *41*:378 (1981).

[AHKTV] S. Albeverio, R. Høegh-Krohn, D. Testard, and A. Vershik, *J. Funct. Anal.*, *51*:115 (1983).

[Alv] O. Alvarez, *Vertex Operators in Mathematics and Physics* (J. Lepowsky, S. Mandelstam, and I. M. Singer, eds.), Publications of the Mathematical Sciences Research Institute, 3, Springer, New York, pp. 37–47 (1984).

[Ar1] H. Araki, *J. Math. Phys.*, *1*:492 (1960).

[Ar2] H. Araki, *Publ. RIMS, Univ. Kyoto*, *5*:361 (1970).

[Arn] V. Arnold, *Ann. Inst. Fourier*, *16*:319 (1966).

[ArW] H. Araki and E.T. Woods, *Publ. RIMS, Kyoto Univ.*, *2*:157 (1966).

[Ba] P. Baxendale, *Comp. Math.*, *53*:19 (1984).

[BaBSS1] F. A. Bais, P. Bouwknegt, M. Surridge, and K. Schoutens, *Nucl. Phys.*, *B 304*:348 (1988).

[BaBSS2] F. A. Bais, P. Bouwknegt, M. Surridge, and K. Schoutens, *Nucl. Phys.*, *B 304*:371 (1988).

[BaMPZ] Yu. A. Bahturin, A. A. Mikhalev, V. M. Petrogradsky, M. V. Zaicev, Infinite dimensional Lie superalgebras, De Gruyter, Berlin (1992).

[Bar] I. Bars, *Vertex Operators in Mathematics and Physics* (J. Lepowsky et al., eds.), MSRI Publ. 3, p. 373 (1983).

[BäuK] G. G. A. Bäuerle, and E. A. de Kerf, Lie Algebras, North Holland, Amsterdam (1990).

[Bax] J. R. J. Baxter, *Exactly Solved Models in Statistical Mechanics*, Academic Press, New York (1982).

[Be] G. Benkart, *Lie Algebras and Related Topics*, CMS Conference 5, pp. 111–138 (1986).

[BoBCM] L. Bonora, M. Bregola, P. Cotta-Ramusino, and M. Martinelli, *Phys. Letts.*, *B205*:53 (1988).

[Bot] R. Bott, *Michigan Math. J.*, *5*:35 (1985).

[BouN] P. Bouwknegt and W. Nahm, *Phys. Lett.*, *B184*:359 (1987).

[Bou] N. Bourbaki, *Groupes et Algèbres de Lie*, CCLS, Paris, Chaps. 1–8 (1957).

[Bouw] P. Bouwknegt, *Nucl. Phys.*, *B290*:507 (1987).

[BPZ] A. Belavin, A. M. Polyakov, and A. B. Zamoldchikov, *Nucl. Phys.*, *B181*:333 (1981).

[Ca1] J. Cardy, *Nucl. Phys.*, *B270*:186 (1986).

[Ca2] J. L. Cardy, *Nucl. Phys.*, *B270*:186 (1986).

[Car] E. Cartan, *Sur la Structure des Groupes Finis et Continus*, Thèse (1894).

[Cart] P. Cartier, *Quantum Mechanical Commutation Relations and Theta Functions*, Proceedings of Symposia in Pure Mathematics, Vol. 9, Amer. Math. Soc., Providence, RI, pp. 361–383 (1966).

[CFRS] R. Coquereaux, L. Frappat, E. Ragoucy, and P. Sorba, *Comm. Math. Phys.*, *133*:1 (1990).

[Che] C. Chevalley, *C. R. Acad. Sci. Paris*, *227*:1136 (1948).

[CiM] R. Cirelli and A. Manià, *J. Math. Phys.* *26*:3036 (1985).

[CIZ] A. Cappelli, C. Itzykson, and J. B. Zuber, *Comm. Math. Phys.*, *112* (1987).

[Co1] A. Connes, *Publ. Math. IHES*, *62*:257 (1986).

[Co2] A. Connes, *C. R. Acad. Sci. Paris*, *296*:953 (1983).

[CoG] F. Constantinescu and H. F. De Groote, Geometrische und Algebraische Methoden der Physik: Supermannigfaltikeiten und Virasoro Algebren, Teubner, Leipzig (1992).

[De1] P. Delorme, *Bull. Soc. Math. France*, *105*:281 (1977).

[De2] P. Delorme, private communication (1988).

[deV] H. J. de Vega, "YBZF Algebras, Kac-Moody Algebras and Integrable Theories," Symposium on Topological and Geometrical Methods in Field Theory, Matinkylä, Espoo, Finland, (1986).

[Dix] J. Dixmier, *Enveloping Algebras*, North-Holland, Amsterdam (1974).

[DJKM] E. Date, M. Jimbo, M. Kashiwara, and T. Miwa, *Publ. RIMS, Kyoto Univ.*, *18*:1077–1119 (1982); *J. Phys. Soc. Japan*, *50*:3306–3318 (1981).

[DJMO] E. Date, M. Jimbo, T. Miwa, and M. Okado, Theta Functions-Bowdoin 1987, Part 1, Proc. Symp. Pure Math., *Amer. Math. Soc.*, Providence, RI, *49*:295–331.

[dlH] P. de la Harpe, *Classical Banach-Lie Algebras and Banach-Lie Groups of Operators on Hilbert Space*, Lecture Notes in Math. vol. 285, Springer, Berlin (1972).

[Do1] L. Dolan, *Phys. Rev. Lett.*, *47*:1371 (1981).

[Do2] L. Dolan, *Phys. Rep.*, *109*:1 (1984).

[Dr] V. Drinfeld, *Quantum Groups*, Proc. Int. Conf. Math. Berkeley (1986).

[Dun1] T. E. Duncan, *Systems Information and Control*, Vol. I (L. R. Hunt and C. F. Martin, eds)., Math. Sci. Press, pp. 123–134 (1984).

[Dun2] T. E. Duncan, *J. Funct. Anal.*, *84*:135 (1989).

[Dyn] E. Dynkin, *Amer. Math. Soc. Transl.*, *17* (1950).

[Fa1] L. D. Faddeev, Braid group, knot theory and statistical mechanics, *World Scient.*, Singapore, pp. 97–110 (1989).

[Fa2] L. D. Faddeev, *Phys. Lett.*, *145 B*:81 (1984).

[FeFröK1] G. Felder, J. Fröhlich, and G. Keller, *Comm. Math Phys.*, *124*:417 (1989).

[FeFröK2] G. Felder, J. Fröhlich, and G. Keller, *Comm. Math Phys.*, *130*:1 (1990).

[FeiFu] B. L. Feigin and D. B. Fuchs, *Funct. Anal. Appl.*, *17*:91 (1983).

[FFr] A. Feingold and I. B. Frenkel, *Adv. Math.*, *56*:117 (1985).

[FiKi] J. M. Figueroa-O'Farrill and T. Kimura, *Comm. Math. Phys.*, *124*:105 (1989).

[FoNi] O. Foda and B. Nienhuis, *Nucl. Phys.*, *B324*:643 (1989).

[Fr1] I. B. Frenkel, *Invent. Math.*, *77*:301 (1984).

[Fr2] I. B. Frenkel, *J. Funct. Anal.*, *44*:259 (1981).

[Fr3] I. B. Frenkel, "Beyond Affine Lie-Algebras," Proc. Int. Conf. Math., vol. 1, Berkeley, pp. 821–839 (1986).

[FrdV] H. Freudenthal and H. de Vries, *Linear Lie Groups*, Academic Press, New York (1969).

[FrGZ] I. B. Frenkel, H. Garland, and G. Zuckerman, *Proc. Natl. Acad. Sci. USA*, *83*:8442 (1986).

[FrK] I. B. Frenkel and V. G. Kac, *Invent. Math.*, *62*:23 (1980).

[FrLM] I. B. Frenkel, J. Lepowsky, and A. Meurman, *Vertex Operator Algebras and the Monster*, Academic Press, Boston (1988).

[FrQS] D. Friedan, Z. Qiu, and S. Shenker, *Vertex Operators in Mathematics and Physics*, MSRI Publ. vol. 3, pp. 419–449 (1984).

[Fu] J. Fuchs, Affine Lie algebras and quantum groups: an introduction with applications in conformal field theory, Cambridge Univ. Press (1992).

[Ga1] H. Garland, *Publ. Math. IHES*, *52*:5 (1980).

[Ga2] H. Garland, *J. Algebra*, *53*:480 (1978).

[GaTr] B. Gaveau and Ph. Trauber, *C. R. Acad. Sci. Paris*, *291A*:575 (1980).

[GeGr] I. M. Gel'fand and M. I. Graev, *Funktional. Anal. Prilozhen*, *2*:20 (1968).

[GeGrPS] I. M. Gelfand, M. I. Graev, and I. I. Pyatetskii-Shapiro, *Representation Theory and Automorphic Functions*, Saunders, Philadelphia (1969).

[GeGrVe] I. M. Gelfand, M. I. Graev, and A. M. Veršik, *Compos. Math.*, *35*:299 (1977); *42*:217 (1981).

[GeGrVi] I. M. Gelfand, M. I. Graev, and N. Ya. Vilenkin, *Generalized Functions*, vol. V, Academic Press, New York (1966).

[GeLiWu] Mo-Lin Ge, Chau Ling-Lie, and Yong-Shi Wu, *Phys. Rev.*, *25D*:1086 (1982).

[GeSh] I. M. Gelfand and G. E. Shilov, *Generalized Functions*, vols. I, II, III, Academic Press, New York (1968).

[GeVi] I. M. Gelfand and N. Y. Vilenkin, *Generalized Functions*, vol. IV, Academic Press, New York (1966).

[GeW] D. Gepner and E. Witten, *Nucl. Phys.*, *B278*:493 (1986).

[GeWu] Mo-Lin Ge and Yong-Shi Wu, *J. Math. Phys.*, *24*:1187 (1983).

[GiWy] L. Girardello and W. Wyss, *Helv. Phys. Acta*, *45*:197 (1972).

[GlJa] J. Glimm and A. Jaffe, *Quantum Physics: A Functional Integral Point of View*, 2nd ed., McGraw-Hill, New York (1987).

[Go] P. Goddard, *Kac Moody and Virasoro Algebras in Mathematical Physics*, Proc. Eighth Int. Congress on Mathematical Physics, Marseille, World Scientific, pp. 390–401 (1987).

[GoKO] P. Goddard, A. Kent, and D. Olive, *Unitary Representations of the Virasoro Algebra and Super-Virasoro Algebras*, Cambridge preprint (1985).

[Gol] V. Ja. Golodec, *Soviet Math. Dokl.*, *9*:184 (1968).

[GoMS1] G. A. Goldin, R. Menikoff, and D. H. Sharp, *J. Phys.*, *16A*:1827 (1983).

[GoMS2] G. A. Goldin, R. Menikoff, and D. H. Sharp, *Phys. Rev. Lett.*, *51*:2246 (1983).

[GoMS3] G. A. Goldin, R. Menikoff, and D. H. Sharp, *Phys. Rev. Lett.*, *58*:2162 (1987).

[GoO1] P. Goddard and D. Olive, *Int. J. Mod. Phys. A*, *1*:303 (1986).

[GoO2] P. Goddard and D. Olive, *Nucl. Phys.*, *257B*:226 (1985).

[GoO3] P. Goddard and D. Olive, *Vertex Operators in Mathematics and Physics* (J. Lepowsky, S. Mardelstam, and I. M. Singer, eds.), Publications of the Mathematical Sciences Research Institute, 3, Springer Verlag, Berlin and New York, pp. 51–96 (1984).

[GoSh1] G. A. Goldin and D. H. Sharp, *Commun. Math. Phys.*, *92*:217 (1983).

[GoSh2] G. A. Goldin and D. H. Sharp, *Phys. Rev.*, *28D*:830 (1983).

[GoWa] R. Goodman and N. R. Wallach, *J. Reine Angew. Math.*, *347*:69 (1984).

[GSW] M. B. Green, I. H. Schwarz, and E. Witten, *Superstring Theory*. I, II, Cambridge University Press, Cambridge (1987).

[Gui] A. Guichardet, *Symmetric Hilbert Spaces and Related Topics*, Lecture Notes in Math., vol. 261, Springer, Berlin (1972).

[Ha1] Z. Haba, *Int. J. Mod. Phys.*, *A4*:267 (1989).

[Ha2] Z. Haba, *J. Math. Phys.*, *32*:19 (1991).

[Hari] Harish-Chandra, *Trans. Amer. Math. Soc.*, *70*:28 (1951).

[He] S. Helgason, *Differential Geometry, Lie Groups and Symmetric Spaces*, 2nd ed., Academic Press, New York (1978).

[Her] Robert Hermann, *Dynamical Systems Defined on Infinite Dimensional Lie Algebras of the "Current Algebra" or "Kac-Moody" Type* (1982).

[Hi] T. Hida, *Brownian Motion*, Springer, Berlin (1980).

[HKT] R. Høegh-Krohn and B. Torrésani, *J. Funct. Anal.*, *89*:106 (1990).

[Is1] R. S. Ismagilov, *Math. USSR Sbornik*, *29*:105 (1976) (transl.).

[Is2] R. S. Ismagilov, *Funct. Anal. Appl.*, *12*:226 (1978).

[Is3] R. S. Ismagilov, *Funct. Anal. Appl.*, *15*:134 (1981).

[JKa] H. P. Jacobsen and V. G. Kac, *Nonlinear Equations in Quantum Field Theory*, Lecture Notes in Physics, vol. 226, Springer, Berlin and New York, pp. 1–20 (1985).

[Jo] P. E. T. Jorgensen, *Operators and Representation Theory*, North-Holland, Amsterdam (1988).

[Ju1] B. Julia, *On Infinite Dimensional Symmetry Groups in Physics*, Proc. Niels Bohr Centennial Conf. p. 215 (1985).

[Ju2] B. Julia, *Vertex Operators in Mathematics and Physics* (J. Lepowsky, S. Mandelstam, and I. Singer, eds.), Springer-Verlag, New York, p. 393 (1983).

[Ka1] V. G. Kac, *Infinite Dimensional Lie Algebras*, 3rd ed., Cambridge University Press (1985).

[Ka2] V. G. Kac (Ed.), *Infinite-Dimensional Groups with Applications*, MSRI Publications, 4 Springer-Verlag, New York (1984).

[Ka3] V. G. Kac, *Math USSR Izv.*, *2*:1271 (1968).

[Ka4] V. G. Kac, *Adv. Math.*, *26*:8 (1977).

[Ka5] V. G. Kac, *Highest Weight Representations of Infinite-Dimensional Lie Algebras*, Proc. Int. Congress Math., Helsinki, pp. 299–304 (1980).

[Kak] M. Kaku, *Introduction to Superstrings*, Graduate Texts in Contemporary Physics, Springer-Verlag (1988).

[Kan] I. L. Kantor, *Soviet Math. Dokl.*, *9*:409 (1968).

[KaP1] V. G. Kac and D. H. Peterson, *Adv. Math.*, *50* (1983).

[KaP2] V. G. Kac and D. H. Peterson, *Proc. Natl. Acad. Sci. USA*, *78*:3308 (1981).

[KaP3] V. G. Kac and D. H. Peterson, *112 Constructions of the Basic Representations of the Loop Group of* E_8, (W. A. Bardeen and A. R. White, eds.), Symp. Anomalies, Geometry, Topology, World Scientific, Singapore, pp. 276–298 (1985).

[KaR] V. G. Kac and A. K. Raina, *Adv. Ser. Math. Phys.*, *2* (1987).

[Kas1] D. Kastler, *$\mathbb{Z}_z$-Graded Cyclic Homology*, Hermann, Paris (1988).

[Kas2] D. Kastler, *Cyclic Cohomology, within the Differential Envelope*, Travaux en cours 30, Hermann (1988).

[KasMaQ] D. Kastor, E. Martinec, and Zongan Qiu, *Phys. Lett.*, *200B*:434–440 (1988).

[KaTo] V. G. Kac and T. Todorov, *Comm. Math. Phys.*, *102*:337 (1985).

[KaW] V. G. Kac and M. Wakimoto, *Adv. Math.*, *70*:156 (1988).

[Kaz] D. A. Kazhdan, *Funct. Anal. Appl.*, *1*:63 (1967) (transl.).

[Ki1] A. N. Kirillov, *Elements of the Theory of Representations*, Springer, Berlin (1978).

[Ki2] A. A. Kirillov, *Sov. Math. Dolk.*, *14*: no. 5 (1973).

[Ki3] A. A. Kirillov, *Sel. Math. Sov.*, *1*:351 (1981).

[Ki4] A. A. Kirillov, *Russ. Math. Surveys*, *22*:63 (1967).

[Kil] W. Killing, *Math. Ann.*, *31*:252 (1888); *33*:1 (1889); *34*:57 (1889); *36*:161 (1890).

[KlS] J. R. Klauder and B. S. Skagerstam, *Coherent States, Applications in Physics*, World Scientific, Singapore (1985).

[KN1] I. M. Krichever and S. P. Novikov, *Funk. Anal. Prilozhen*, *21–2*:46 (1987).

[KN2] I. M. Krichever and S. P. Novikov, *Funk. Anal. Prilozhen*, *21–4*:47 (1987).

[KoRo] W. Kondracki and J. Rogulski, Dissertationes Math. (Rozprawy Mat.) *150*:1 (1986).

[Ku] S. Kusuoka, Ideas and Methods in Mathematical Analysis, Stochastics, and Applications, In Memory of Raphael Høegh-Krohn, Vol. 1 (S. Albeverio, J. E. Fenstad, H. Holden, and T. Lindstrøm, eds.), Cambridge University Press, Cambridge, pp. 405–424 (1992).

[Kul] P. P. Kulish, Ed., *Quantum Groups*, Lect. Notes Math., vol. 1510, Springer, Berlin (1992).

[KZ] V. G. Knizhnik and A. B. Zamolodchikov, *Nucl. Phys.*, *B247*:83 (1984).

[Le1] J. Lepowsky, *Lie Algebras and Combinatorics*, Proc. ICM, Helsinki (1978).

[Le2] J. Lepowsky, *Vertex Operators in Mathematics and Physics* (J. Lepowsky, S. Mandelstam, and I. M. Singer, eds.), Publications of the Mathematical Sciences Research Institute, 3, Springer, pp. 1–13 (1984).

[LeW] J. Lepowsky and R. L. Wilson, *Adv. Math.*, *45*:21 (1982).

[Lu] G. Lusztig, *Astérisque*, *101–102*:208 (1983).

[Ma1] J. Marion, *J. Funct. Anal.*, *54*:1 (1983).

[Ma2] J. Marion, *Nouvelles Classes d'Algèbres de Lie Affines*, Marseille preprint (1987).

[Ma3] J. Marion, *Exp. Math.*, *6*:177 (1988).

[Ma4] J. Marion, *Unitarisation*—process for some nonlocally compact topological groups, to appear in *Math. Zeitschr.* (1993).

[MaEbFi] J. Marsden, D. Ebin, and A. Fisher, *Diffeomorphism Groups, Hydrodynamics and Relativity* (J. F. Vanstone, ed.), Proc. 13th Biennal Seminar Can. Math. Congress, Montreal, pp. 135–279 (1972).

[Man] S. Mandelstam, *Vertex Operators in Mathematics and Physics* (J. Leposwky, S. Mandelstam, and I. M. Singer, eds.), Publications of the Mathematical Sciences Research Institute, 3, Springer, pp. 15–33 (1984).

[Man] Y. I. Manin, Topics in noncommutative geometry, Princeton University Press (1991).

[Mat] O. Mathieu, *C. R. Acad. Sci. Paris, Sér. I*, *306*:227 (1988).

[MaT] J. Marion and D. Testard, *J. Funct. Anal.*, *76*:160 (1988).

[Math] D. Mathon, *Proc. Cambridge Philos. Soc.*, *72*:347 (1972).

[MaW] J. Marsden and A. Weinstein, *Rep. Math. Phys.*, *5*:121 (1974).

[McD] I. G. MacDonald, *Invent. Math.*, *15*:91 (1972).

[McK] J. McKenna and J. R. Klauder, *J. Math. Phys.*, *5*: no. 7 (1964).

[Mi1] J. Mickelsson, *Phys. Rev.*, *D32*:436 (1985).

[Mi2] J. Mickelsson, *Comm. Math. Phys.*, *97*:361 (1985).

[Mi3] J. Mickelsson, *J. Math. Phys.*, *26*:377 (1985).

[Mi4] J. Mickelsson, Current Algebras and Groups, Plenum Press, New York (1989).

[Mic] P. W. Michor, *Cahiers Topologie Geometrie Diff.*, *24*:57 (1984).

[Mil] J. Milnor, "Les Houches" (B. S. DeWitt and R. Stora, eds.), Session XL, 1983, Relativité, Groupes et Topologie II, Elsevier (1984).

[Mir] S. Miracle-Sole, *Ann. Inst. H. Poincaré, Sect. A*, *6*:59 (1967).

[MiRa] J. Mickelsson and S. G. Rajeev, *Comm. Math. Phys.*, *116*:365 (1988)

[Moo1] R. V. Moody, *J. Algebra*, *10*:211 (1968).

[Moo2] R. V. Moody, *Canad. J. Math.*, *21*:1432 (1969).
[Mor1] J. Morita, *Tsukuba J. Math.*, *6*:1 (1982).
[Mor2] J. Morita, *Comm. Alg.*, *12*:673 (1984).
[Mor3] J. Morita, *Hokkaido Math. J.*, *13*:51 (1984).
[Mor4] J. Morita, *Proc. Japan Acad. Ser. A Math. Sci.*, *63*:21 (1987).
[Na] K. S. Narain, *Phys. Lett.*, *169B*:41 (1986).
[Ne] E. Nelson, *J. Funct. Anal.*, *12*:211 (1973).
[Ner] Yu. A. Neretin, *Adv. Sov. Maths.*, *2*:103 (1991) (transl.).
[O] E. P. Osipov, *Lett. Math. Phys.*, *18*:35 (1988).
[OkS] K. Okamoto, T. Sakurai, *Hiroshima Math. J.*, *12*:529 (1982).
[Ol] G. I. Ol'shanskii, *Funct. Anal. Appl.*, *12*:185 (1978).
[Oli] D. Olive, *Infinite-Dimensional Algebras in Modern Theoretical Physics*, Proc. Eighth Int. Congress on Mathematical Physics, Marseille, World Scientific, pp. 242–256 (1987).
[Om] H. Omori, *Infinite Dimensional Lie Transformations Groups*, Lecture Notes in Math., vol. 427, Springer, Berlin (1974).
[Or] B. Ørsted, *J. Funct. Anal.*, *36*:53 (1980).
[P] D. H. Peterson, *Affine Lie Algebras and Theta Functions*, Lecture Notes in Math., vol. 933, Springer, Berlin (1982).
[Pa] V. Pasquier, *Nucl. Phys.*, *B295*:491 (1988).
[PaS1] K. Parthasarathy and K. Schmidt, *Acta Math.*, *128*:53 (1971).
[PaS2] K. Parthasarathy and K. Schmidt, Positive definite kernels, continuous tensor products and central limit theorems of probability theory, *Lecture Notes in Math., vol. 272*, Springer, Berlin (1972).
[Pe] A. Perelomov, *Generalized Coherent States and Their Applications*, Springer, Berlin (1986).
[Pi1] D. Pickrell, *Proc. Amer Math Soc.*, *102*:416 (1988).
[Pi2] D. Pickrell, *J. Funct. Anal.*, *90*:1 (1990).
[PrS] A. Pressley and G. Segal, *Loop Groups*, Clarendon Press, Oxford (1986).
[PT] R. R. Paunov and I. T. Todorov, *Local Extensions of the U(1) Current Algebra and Their Positive Energy Representations* (V. G. Kac, ed.), Proc. Conf. Infinite-Dimensional Lie Algebras and Lie Groups, CIRM Marseille Luminy (1988).
[Pu] C. R. Putnam, *Commutation Properties of Hilbert Space Operators and Related Topics*, Springer, Berlin (1967).
[Rag] M. S. Raghunathan, *Kirillov Theory*, College on Representation Theory of Lie Groups, Trieste (1985).
[Rav] F. Ravanini, *Mod. Phys. Lett.*, *A3*:397 (1988).
[Ree] M. Reeken, *Stromkommutatoren, Schwingerterme und die Jacobi-Identität im freien Diracfeld*, Dissertation, Wien (1968).
[ReSe] A. Reiman and M. Semenov-Tian-Shanskii, *Inv. Math.*, *54*:81 (1979); *63*:423 (1981).
[ReSi] M. Reed and B. Simon, *Methods of Modern Mathematical Physics*. I–IV, Academic Press, London (1980).

[Rit] V. Rittenberg, Conformal Invariance and String Theory, Academic Press, Boston, pp. 37–61 (1989).
[Rö] M. Röckner, *J. Funct. Anal.*, *79*:211 (1988).
[Ros] M. Rosso, *Comm. Math. Phys.*, *117*:581 (1988).
[Sa] M. Sato, *RIMS Kokyuoku*, *439*:30 (1981).
[San] N. Sanchez, *Semiclassical Quantum Gravity: Recent Results in Two and Four Dimensions* (A. Jadczyk, ed.), Proc. XXII Karpacz Winter School and Workshop of Theoretical Physics, World Scientific (1986).
[SaVe] M. V. Saveliev and A. M. Vershik, *Comm. Math. Phys.*, *126*:367 (1989).
[Sch] B. Schroer, Differential geometrical methods in theoretical physics, Kluwer, Dordrecht, pp. 219 (1988).
[ScWa] A. N. Schellekens and N. P. Warner, *Phys. Rev.*, *D34*:3092 (1986).
[Se1] G. B. Segal, *Loop Groups*, Lecture Notes in Math., vol. 1111, F. Hirzebruch, ed., Springer, Berlin, pp. 155–168 (1985).
[Se2] G. B. Segal, *Comm. Math. Phys.*, *80*:301 (1981).
[Seg] I. E. Segal, *Ann. Math.*, *66*:297 (1957).
[SeW] G. Segal and G. Wilson, *Publ. Math. IHES*, *61*: (1985).
[Sh] E. T. Shavgulidze, *Funct. Anal. Appl.*, *12*:203 (1978).
[Sha] I. Shafarevich, *Math. USSR Izv.*, *18*:214 (1982).
[Si] B. Simon, *The $P(\phi)_2$ Euclidian (Quantum) Field Theory*, Princeton University Press, Princeton, NJ (1974).
[Sin] I. M. Singer, Proc. Conf. Elie Cartan et les Mathématiques d'Aujourd'hui, Lyon (1986).
[Sk] T. H. R. Skyrme, *J. Math. Phys.*, *12*:1735 (1971).
[Sl] P. Slodowy, Publ. *RIMS*, *415*:19 (1981).
[Sm] S. Smale, *J. Math. Mech.*, *14*:315 (1965).
[SoT] P. Sorba and B. Torrésani, *Int. J. Mod. Phys.*, *A3*:1451 (1988).
[SpSeTroVP] Ph. Spindal, A. Sevrin, W. Troost, and A. Van Proyen, *Nucl. Phys.*, *B 308*:663 (1988).
[St] A. V. Štraus, *Soviet Math. Dokl.*, *9*:205 (1968).
[Sto] D. T. Stoyanov, Conformal groups and related symmetries: physical results and mathematical background, Lect. Notes Phys., Springer, Berlin, pp. 379–386 (1986).
[Str1] R. F. Streater, *Nuovo Cimento*, *A53*:487 (1968).
[Str2] R. F. Streater, *Rend. Sti. Isc. Fis. E. Fermi*, *11*:247 (1969).
[Str3] R. F. Streater, *Z. Wahrsch. Gebiete*, *19*:67 (1971).
[StrV] S. Strătilă, D. Voiculescu, Representations of AF-Algebras and the group $\Omega(\infty)$, Lect. Notes Math. 486, Springer, Berlin (1975).
[Te] D. Testard, *Representations of the Group of Equivariant Loops in SU(n)*, Bielefeld preprint (1985).
[Th-M] J. Thierry-Mieg, *Non-Perturbative Quantum Field Theory* ('t Hooft et al., eds.), Cargese Lecture Notes.

[Ti] J. Tits, *Groups and Group Functions Attached to Kac-Moody Data*, Lecture Notes in Math., vol. 1111, Springer, Berlin, pp. 193–223 (1985).

[To1] B. Torrésani, *Unitary Highest Weight Representations of Gauge Groups*, to appear in Memorial Volume for R. Høegh-Krohn.

[To2] B. Torrésani, *Représentations Projectives des Groupes de Transformations de Jauge Locales*, Ph.D. thesis, Marseille (1986).

[To3] B. Torrésani, *Lett. Math. Phys.*, *13*:7 (1987).

[Tod] I. T. Todorov, *Infinite Dimensional Lie Algebras in Conformal QFT Models* (A. O. Barut, ed.), Lecture Notes in Physics, vol. 261, Springer, Berlin, pp. 387–443 (1986).

[Tr] S. Treiman, R. Jackiw, B. Zumino, and E. Witten, *Current Algebras and Anomalies*, World Scientific, Singapore (1985).

[TzWu] Chia-Hsiung Tze and Yong-Shi Wu, Nucl. Phys. *B204*:118 (1982).

[VZ] A. M. Vershik and D. P. Zhelobenko, Eds., Representation of Lie groups and related topics, Gordon and Breach, New York (1990).

[Va] V. S. Varadarajan, *Lie Groups, Lie Algebras and Their Representations*, Springer, Berlin (1984).

[VeGeGr1] A. M. Vershik, I. M. Gelfand, and M. I. Graev, *Russ. Math. Surv.*, *28*:83 (1973).

[VeGeGr2] A. M. Vershik, I. M. Gelfand, and M. I. Graev, *Funct. Anal. Appl.*, *8*:151 (1974) (transl.).

[VeGeGr3] A. M. Vershik, I. M. Gelfand, and M. I. Graev, *Compos. Math.*, *42*:217 (1981).

[VeK] A. M. Vershik and S. I. Karpushev, *Math. USSR Sbornik*, *47*:513 (1984) (transl.).

[Ver] J. L. Verdier, *Sém. Bourbaki, Astér.*, *92–93*:365 (1983).

[Verm] D. N. Verma, *Bull. Amer. Math. Soc.*, *74*:160 (1968).

[Wa] N. Wallach, *Comp. Math.*, *64*:3 (1987).

[Wan] S. P. Wang, *Proc. Amer. Math. Soc.*, *42*:621 (1974).

[Wi] R. Wilson, *Euclidean Lie Algebras Are Universal Central Extensions*, Lecture Notes in Math., vol. 933, Springer, Berlin, pp. 210–213 (1982).

[WiGa] A. Wightman and L. Gårding, *Ark. Fysik*, *28*:129 (1965).

[Wo] S. Woronowicz, Comm. Math. Phys. *122*:125 (1989).

[Wu1] Yong-Shi Wu, *Phys. Lett.*, *96A*:179 (1983).

[Wu2] Yong-Shi Wu, *Comm. Math. Phys.*, *90*:461 (1983).

[Za] W. J. Zakrzewski, *Low Dimensional Sigma Models*, Adam Hilger, Bristol (1989).

[Zam] A. B. Zamolodchikov, *Th. Math. Phys.*, *65*:1205 (1985).

[Xi] D. X. Xia, Measure and Integration Theory on Infinite-Dimensional Spaces, Academic Press, New York (1972).

1

Basic Functional Groups and Lie Algebras

Many problems of classical functional analysis are formulated in the setting of Hilbertian triads ($\mathcal{D}(X)$, $W^{(k)}_{2,\mu}(X)$, $\mathcal{D}'(X)$) and some closely related spaces, where X is some Riemannian manifold endowed with a positive measure μ; $\mathcal{D}(X)$ denotes the space of test functions on X equipped with the Schwartz topology (which makes it a nuclear space, countable strict inductive limit of Fréchet spaces), $\mathcal{D}'(X)$ denotes the space of distributions on X, and $W^{(k)}_{2,\mu}(X)$ denotes the Hilbert-Sobolev completion of $\mathcal{D}(X)$ with respect to the Sobolev norm:

$$f \to \|f\|^{(k)}_{2,\mu} = \left[\int_X \sum_{|\alpha| \leq k} \left|\frac{\partial^{|\alpha|} f}{\partial x_1^{\alpha_1} \cdots \partial x_n^{\alpha_n}}(x)\right|^2 d\mu(x)\right]^{1/2}, \qquad k \in \mathbb{N}$$

Also useful are Hilbertian triads corresponding to Hilbert-valued test functions: ($\mathcal{D}(X) \otimes \mathcal{H}$, $W^{(k)}_{2,\mu}(X) \otimes \mathcal{H}$, $\mathcal{D}'(X) \otimes \mathcal{H}$), with $\mathcal{H}$ a separable Hilbert space.

From a "differential" and "group-theoretic" point of view, the real finite-dimensional Hilbert spaces $\mathcal{H}$ are the simplest real Lie groups. So, it is natural to consider, instead of $\mathcal{H}$-valued test functions on X, G-valued test functions, G being any real Lie group. In the following we shall discuss precisely such a noncommutative extension.

Let X be a Riemannian manifold, let G be a real Lie group with Lie algebra $\mathcal{G}$, and let $\mathcal{D}(X, G)$ be the set of G-valued C^∞-functions on X which are compactly supported, i.e., taking as value the unit element of G outside a compact subset of X. We can endow $\mathcal{D}(X, G)$ with the group structure coming from G by the pointwise product.

We endow $\mathcal{D}(X, G)$ with an analogue of the Schwartz topology for which it becomes a topological group having some properties analogous to those of Lie groups and nuclear spaces. Thus we assume there exists a Lie algebra $\mathcal{L}$ which, as vector space, is the test function space belonging to some Hilbertian triad, and such that there exists some neighborhood of the unit element of $\mathcal{D}(X, G)$ homeomorphic to a neighborhood of the neutral element (zero) of $\mathcal{L}$.

$\mathcal{D}(X) \otimes \mathcal{G}$, endowed with the pointwise algebraic structures coming from $\mathcal{G}$, and with the Schwartz topology, seems to be a good candidate for $\mathcal{L}$. In fact this chapter is mainly devoted to the goal of endowing $\mathcal{D}(X, G)$ with a nuclear Lie group structure in the sense of [1, Chapter 4, Section 5-4]. We shall show that the Hilbert-Sobolev completion $W^{(k)}_{2,\mu}(X) \otimes \mathcal{G}$ of $\mathcal{D}(X, G)$, under some conditions involving k and the dimension of X, is the Hilbert-Sobolev Lie algebra of some Hilbert Lie group $W^{(k)}_2(X, G)$ which is the completion of $\mathcal{D}(X, G)$ with respect to some Sobolev norm.

In the case $k = 1$, and for G being a compact semisimple Lie group, we give, in this chapter, another way to get a complete metric group $H^1_q(X, G)$, $q \geq 1$, with respect to a metric depending on the first derivative of elements of $\mathcal{D}(X, G)$, without involving conditions on $\dim(X)$, by using the so-called Maurer-Cartan cocycle and of a positive definite quadratic form on the space $\Omega(X)$ of compactly supported smooth 1-forms on X. Moreover, we give some information about some important subgroups of $\mathcal{D}(X, G)$ and about jet groups.

1.1 NOTATIONS AND PRELIMINARIES

Let G denote a finite-dimensional and connected real Lie group. Its Lie algebra $\mathcal{G}$ is taken to be a subalgebra of the Lie algebra $\mathcal{L}(V)$ of all the endomorphisms of some finite-dimensional real vector space V, the bracket law being, as usual, given by $[u, v] = u \cdot v - v \cdot u$. We shall denote by s the dimension of $\mathcal{G}$, by e the unit element of G, and by 0 the neutral element of $\mathcal{G}$.

We equip $\mathcal{L}(V)$ with a scalar product $\langle\ ,\ \rangle$, and without loss of generality, we suppose that the corresponding Euclidean norm $|\ |$ is admissible on $\mathcal{G}$, i.e., such that for all u and v in $\mathcal{G}$,

$$|[u, v]| \leq |u| \cdot |v| \tag{1.1}$$

If r is a strictly positive real number, the open ball with radius r and center at the origin in $\mathcal{G}$ will be denoted by B_r: $B_r = \{u \in \mathcal{G} \mid |u| < r\}$.

X denotes a Riemannian manifold with finite dimension n endowed with a volume measure dx. We select a sequence $(K_i)_i$ of compact subsets of X such that (with $\mathring{K}_i$ the interior of K_i):

$$\mathring{K}_0 \neq \emptyset$$

$$K_i \subset \mathring{K}_{i+1} \qquad \text{for all integers } i \tag{1.2}$$

$$X = \bigcup_{i=0}^{i=\infty} K_i$$

We shall denote by $K(X)$ the family $(K_i)_i$. We select also a countable family (O_p, φ_p) of local charts which covers X, where φ_p is a diffeomorphism from the open subset O_p of X onto some open subset of $\mathbb{R}^n$.

Let $\alpha = (\alpha_1, \ldots, \alpha_n)$ be an element of $\mathbb{N}^n$, where $\mathbb{N}$ denotes the set of nonnegative integers. We set $|\alpha| = \alpha_1 + \cdots + \alpha_n$ and $\alpha! = \alpha_1!\alpha_2! \cdots !\alpha_n!$; if $\beta = (\beta_1, \ldots, \beta_n)$ is another element of $\mathbb{N}^n$ we shall write $\beta \leq \alpha$ iff $\beta_i \leq \alpha_i$, $i = 1, \ldots, n$, and in this case we shall denote by $\binom{\alpha}{\beta}$ the number $\alpha!/(\beta!(\alpha - \beta)!)$.

We select an orthonormal basis $a_1, \ldots, a_s$ of $\mathcal{G}$. To each k in $\mathbb{N}$ there corresponds the set

$$A(n, k) = \{\alpha \in \mathbb{N}^n \mid |\alpha| \leq k\}.$$

Let U be an element of $C^k(X, \mathcal{G})$; $U \circ \varphi_p^{-1}$ is a C^k-mapping from the open subset $\varphi_p(O_p)$ of $\mathbb{R}^n$ into $\mathcal{G}$ which can be written

$$U \circ \varphi_p^{-1} = \sum_{i=1}^{i=s} U_i^p(\cdot) a_i,$$

where the U_i^p are elements of the set $C^k(\varphi_p(O_p))$. For α in $A(n, k)$ we shall denote by $\partial^\alpha(U \circ \varphi_p^{-1})$ the mapping from $\varphi_p(O_p)$ into $\mathcal{G}$ given by

$$\partial^\alpha(U \circ \varphi_p^{-1})(x) = \sum_{i=1}^{i=s} \frac{\partial^{|\alpha|} U_i^p(x)}{\partial x_1^{\alpha_1} \cdots \partial x_n^{\alpha_n}} a_i$$

which is the expression of the derivative of U according to α in the local chart (O_p, φ_p). We shall denote by $[U]$ the support of U.

1.2 THE LIE ALGEBRAS $C_0^k(X, \mathcal{G})$ AND $\mathcal{D}(X, \mathcal{G})$

1.2.1 The Fréchet Lie Algebra $C_0^k(X, \mathcal{G})$

Let k be in $\mathbb{N}$, and for all K_i in $K(X)$ let us consider the real vector space $C^k_{K_i}(X, \mathcal{G})$ consisting of all the mappings of class C^k from X into $\mathcal{G}$ with support in K_i. For all real numbers λ and for all pairs (U, V) of elements of $C^k_{K_i}(X, \mathcal{G})$, the mappings

$$U + V: x \to U(x) + V(x), \qquad x \in X$$
$$\lambda U: x \to \lambda U(x), \qquad x \in X$$

have their supports in K_i and are of class C^k. Moreover, let us consider the mapping

$$[U, V]: x \to [U(x), V(x)] = U(x) \cdot V(x) - V(x) \cdot U(x)$$

According to the Leibniz rule of derivation, for α in $A(n, k)$, on the open subset $\varphi(O_p)$ of $\mathbb{R}^n$ one has

$$\partial^\alpha([U, V] \circ \varphi_p^{-1}) = \sum_{0 \leq \beta \leq \alpha} \binom{\alpha}{\beta} [\partial^\beta(U \circ \varphi_p^{-1}), \partial^{\alpha-\beta}(V \circ \varphi_p^{-1})]$$

from which it follows that $[U, V]$ is in $C^k_{K_i}(X, \mathcal{G})$. It follows that $C^k_{K_i}(X, \mathcal{G})$, endowed with the pointwise algebraic structure coming from the Lie algebra structure of $\mathcal{G}$, is a real Lie algebra.

Let us consider the space $C_0^k(X, \mathcal{G})$ consisting of all the mappings of class C^k from X into $\mathcal{G}$ with compact support; from (1.2) one has

$$C_0^k(X, \mathcal{G}) = \bigcup_{K_i \in K(X)} C^k_{K_i}(X, \mathcal{G}) \tag{1.3}$$

For any element U of $C^k_{K_i}(X, \mathcal{G})$ the number

$$|U|_{k,i} = \max_{\alpha \in A(n,k)} \left\{ \sup_p \left[\sup_{x \in \varphi_p(O_p \cap K_i)} |\partial^\alpha(U \circ \varphi_p^{-1})(x)| \right] \right\} \tag{1.4}$$

is well defined. As it is well-known, $|\ |_{k,i}$ is a norm on $C^k_{K_i}(X, \mathcal{G})$ for which it is a real Banach space. As $i \leq j$ implies $K_i \subset K_j$, and $C^k_{K_i}(X, \mathcal{G}) \subseteq C^k_{K_j}(X, \mathcal{G})$, it follows from (1.3) that $C_0^k(X, \mathcal{G})$ is a real Fréchet space for which the family $(|\ |_{k,i})_i$ is an increasing sequence of seminorms.

Let $\mathcal{E}$ be the set of decreasing sequences $\varepsilon = (\varepsilon_i)_i$ of strictly positive real numbers ε_i such that $\lim_{i \to +\infty} \varepsilon_i = 0$. As it is well known, the family of sets of the form $\mathcal{V}_k(\varepsilon) = \{U \in C_0^k(X, \mathcal{G}) / |U|_{k,i} < \varepsilon_i, \forall\, i \in \mathbb{N}\}$, where

ε is in $\mathscr{E}$, is a fundamental system of neighborhoods of O in $C_0^k(X, \mathscr{G})$ [2, Chapter 3].

We verify that this topology is compatible with the Lie algebra structure of $C_0^k(X, \mathscr{G})$. For this it suffices to prove that a neighborhood $\mathscr{V}_k(\varepsilon)$ of O being given, we can find ε^1 and ε^2 in $\mathscr{E}$ such that for all U in $\mathscr{V}_k(\varepsilon^1)$ and all V in $\mathscr{V}_k(\varepsilon^2)$, $[U, V]$ belongs to $\mathscr{V}_k(\varepsilon)$. Let U and V be in $C_0^k(X, \mathscr{G})$; from (1.4) and the Leibniz rule of derivation, one has

$$|[U, V]|_{k,i} = \max_{\alpha \in A(n,k)} \left\{ \sup_p \sup_{x \in \varphi_p(O_p \cap K_i)} \left[\sum_{0 \le \beta \le \alpha} \binom{\alpha}{\beta} \times [\partial^\beta(U \circ \varphi_p^{-1})(x), \partial^{\alpha-\beta}(V \circ \varphi_p^{-1})(x)] \right] \right\}$$

from which it follows that, taking into account (1.1),

$$|[U, V]|_{k,i} \le \lambda \, |U|_{k,i} \cdot |V|_{k,i},$$

where λ denotes the number $\max_{\alpha \in A(n,k)} [\sum_{0 \le \beta \le \alpha} \binom{\alpha}{\beta}]$. In order to get what we wish it suffices to take ε^1 and ε^2 such that $\varepsilon_i^1 = \sqrt{\varepsilon_i/\lambda}$ and $\varepsilon_i^2 = \sqrt{\varepsilon_i/\lambda}$, $i \in \mathbb{N}$. From the above discussion it follows that the following proposition holds:

Proposition 1. For all k in $\mathbb{N}$, $C_0^k(X, \mathscr{G})$ endowed with the pointwise algebraic structure coming from the Lie algebra structure of $\mathscr{G}$ and with the family of seminorms $(| \, |_{k,i})_i$ is a Fréchet-Lie algebra.

1.2.2 The LF-Nuclear Lie Algebra $C_0^\infty(X, \mathscr{G}) = \mathscr{D}(X, \mathscr{G})$

Let us consider now the space of all the $\mathscr{G}$-valued and compactly supported C^∞-mappings on X, i.e.,

$$C_0^\infty(X, \mathscr{G}) = \bigcap_{k \in \mathbb{N}} C_0^k(X, \mathscr{G}) \tag{1.5}$$

It follows from the above proposition that $C_0^\infty(X, \mathscr{G})$ endowed with the pointwise algebraic structure coming from the Lie algebra structure of $\mathscr{G}$ is a Lie algebra. Let us first consider the case where $\mathscr{G}$ is the field $\mathbb{R}$ of real numbers. As it is well known the Schwartz topology of $C_0^\infty(X, \mathbb{R})$ endows this space with a topology of topological vector space which is separable, locally convex, nuclear, and which is exactly the strictly inductive limit of the topologies of the Fréchet spaces $C_0^k(X, \mathbb{R})$ (see, e.g., [2] and [3]); in the usual terminology $C_0^\infty(X, \mathbb{R})$ is called a real LF-nuclear space.

We return now to the general case $C_0^\infty(X, \mathcal{G})$; for all integers k one has $C_0^k(X, \mathcal{G}) = C_0^k(X, \mathbb{R}) \otimes \mathcal{G}$, and $C_0^\infty(X, \mathcal{G}) = C_0^\infty(X, \mathbb{R}) \otimes \mathcal{G}$.

We can then endow $C_0^\infty(X, \mathcal{G})$ with the Schwartz topology coming from that of $C_0^\infty(X, \mathbb{R})$ according to the tensor product by $\mathcal{G}$. On the other hand, the topology of Fréchet space that we have defined on $C_0^k(X, \mathcal{G})$ is exactly that of $C_0^k(X, \mathbb{R}) \otimes \mathcal{G}$. Moreover, similarly as in the proof of Proposition 1, it follows that the bracket law in $C_0^\infty(X, \mathcal{G})$ is compatible with the strictly inductive limit of the topologies of the $C_0^k(X, \mathcal{G})$, k in $\mathbb{N}$. Thus we have

Proposition 2. $C_0^\infty(X, \mathcal{G})$ endowed with the pointwise Lie algebra structure coming from that of $\mathcal{G}$ and with the strictly inductive limit of topologies of the Fréchet-Lie algebras $C_0^k(X, \mathcal{G})$, for all k in $\mathbb{N}$ is a LF and nuclear real Lie algebra.

We denote the nuclear real Lie algebra given by Proposition 2 by $\mathcal{D}(X, \mathcal{G})$. $\mathcal{D}(X, \mathcal{G})$ can be looked upon as the space of $\mathcal{G}$-valued test functions on X.

REMARK. $C_0^k(X, \mathcal{G})$, and $C_0^\infty(X, \mathcal{G})$, from the topological vector space point of view, are well-known spaces; in particular their topologies do not depend on the choice of the family $\mathcal{K}(X) = (K_i)_i$ and the collection of local charts $(O_p, \varphi_p)_p$ (see, e.g., [2, Chapter 3] or [3, Chapter 17]).

1.3 THE HILBERT-SOBOLEV LIE ALGEBRAS $W_2^{(k)}(X, \mathcal{G})$

We recall that X is endowed with a volume measure dx. For all U in $\mathcal{D}(X, \mathcal{G})$, let $\|U\|_2$ be the number

$$\|U\|_2 = \left[\int_X |U(x)|^2 \, dx\right]^{1/2} \tag{1.6}$$

Using the local charts $(O_p, \varphi_p)_p$, on each O_p we can find a smooth density ρ_p such that

$$\|U\|_2 = \sum_p \left[\int_{O_p} |(U \circ \varphi_p^{-1})(y)|^2 \, \rho_p(y) \, d_p(y)\right]^{1/2}$$

when $d_p(y)$ is the Lebesgue measure on O_p.

To each positive integer k there corresponds the Hilbert-Sobolev norm $\| \ \|_2^{(k)}$ on $\mathcal{D}(X, \mathcal{G})$ such that, for all elements U in $\mathcal{D}(X, \mathcal{G})$,

$$\|U\|_2^{(k)} = \left[\sum_{\alpha \in A(n,k)} \left\{\sum_p \int_{O_p} |\partial^\alpha (U \circ \varphi_p^{-1})(y)|^2 \, \rho_p(y) \, d_p(y)\right\}\right]^{1/2} \tag{1.7}$$

The real Hilbert space spanned by $\mathcal{D}(X, \mathcal{G})$ with respect to the Euclidean norm $\| \|_2^{(k)}$ is called the *Hilbert-Sobolev space of order k* spanned by $\mathcal{D}(X, \mathcal{G})$, and is denoted by $W_2^{(k)}(X, \mathcal{G})$.

From the Sobolev vector space point of view, the following results are well known (see, e.g., [4, Chapter 1, Section 8-10]):

If $k > (1/2) \dim (X)$, there exists $M_k > 0$ such that

$$\|U\|_\infty = \sup_{x \in X} |U(x)| \leq M_k \|U\|_2^{(k)} \qquad \text{for all } U \text{ in } \mathcal{D}(X, \mathcal{G}) \tag{1.8}$$

Let U be in $\mathcal{D}(X, \mathcal{G})$;

$$\text{for all } \alpha \text{ in } A\left(n, k - \frac{n}{2}\right) \qquad \partial^\alpha (U \circ \varphi_p^{-1}) \text{ is continuous.} \tag{1.9}$$

We shall now discuss the question whether $W_2^{(k)}(X, \mathcal{G})$ is a Hilbert-Lie algebra for the extended Lie algebra structure coming from that of $\mathcal{D}(X, \mathcal{G})$, that is to say whether there exists a real number $N_k > 0$ such that for all U, V in $\mathcal{D}(X, \mathcal{G})$,

$$\|[U, V]\|_2^{(k)} \leq N_k \|U\|_2^{(k)} |V\|_2^{(k)}$$

It suffices to study this problem by taking for X a bounded open subset of $\mathbb{R}^n$; in this case, for U in $\mathcal{D}(X, \mathcal{G})$,

$$\|U\|_2^{(k)} = \left[\sum_{\alpha \in A(n,k)} \int_X |\partial^\alpha U(x)|^2 \rho(x) \, d_n(x) \right]^{1/2}$$

where $d_n x$ is the Lebesgue measure on X in $\mathbb{R}^n$, and ρ some strictly positive C^∞-density on X. Then

$$\|U\|_2^{(k)} = \left[\sum_{\alpha \in A(n,k)} \|\partial^\alpha U\|_2^2 \right]^{1/2}$$

where $\| \|_2$ denotes the L^2-norm:

$$\|V\|_2 = \left[\int_X |V(x)|^2 \rho(x) \, d_n(x) \right]^{1/2}$$

We have the following result.

Proposition 3. If $k > (1/2) \dim (X)$, there exists $N_k > 0$ such that for all U, V in $\mathcal{D}(X, \mathcal{G})$:

$$\|[U, V]\|_2^{(k)} \leq N_k \|U\|_2^{(k)} \cdot \|V\|_2^{(k)} \tag{*}$$

In particular $W_2^{(k)}(X, \mathcal{G})$ is a Hilbert-Lie algebra.

PROOF. As vector space $W_2^{(k)}(X, \mathcal{G}) = W_2^{(k)}(X, \mathbb{R}) \otimes \mathcal{G}$; moreover it is known that, as soon as $k > (1/2)\dim(X)$, $W_2^{(k)}(X, \mathbb{R})$ endowed with the pointwise product of functions is a Hilbert algebra [5, Corollary 9.7]; thus, if $k > (1/2)\dim(X)$, there exists $\alpha > 0$ such that, for all f_1, f_2 in $W_2^{(k)}(X, \mathbb{R})$:

$$\|f_1 \cdot f_2\|_2^{(k)} \leq \alpha \|f_1\|_2^{(k)} \|f_2\|_2^{(k)} \tag{**}$$

Let $(H_q)_q$ be an orthonormal basis for $\mathcal{G}$ with respect to the scalar product $\langle\ ,\ \rangle$, and let $U = \Sigma_q f_q \otimes H_q$ and $U' = \Sigma_q f'_q \otimes H_q$ be two elements in $W_2^{(k)}(X, \mathcal{G})$. For each pair (q, q'), $1 \leq q \leq \dim(\mathcal{G})$, $1 \leq q' \leq \dim(\mathcal{G})$, there exists a sequence $(a_j(q, q'))_j$ in l_2 such that $[H_q, H_{q'}] = \Sigma_j a_j(q, q')H_j$, from which it follows that

$$[U, U'] = \sum_j \left(\sum_{q,q'} a_j(q, q') f_q \cdot f_{q'} \right) \otimes H_j$$

The assertion then follows immediately from (**). □

We shall call the Hilbert-Lie algebra $W_2^{(k)}(X, \mathcal{G})$ discussed in Proposition 3 the *Hilbert-Sobolev Lie algebra of order k*.

1.4 TECHNICAL RESULTS ABOUT SOME CONTINUOUS SUMS OF OPERATORS

Let $\mathcal{L}(\mathcal{G})$ be the space of all the linear mappings from $\mathcal{G}$ into $\mathcal{G}$, and let A be an element of $C^\infty(X, \mathcal{L}(\mathcal{G}))$. A gives rise to an operator $\tilde{A}$ from $\mathcal{D}(X, \mathcal{G})$ into the space $\mathcal{A}_0(X, \mathcal{G})$ of all $\mathcal{G}$-valued compactly supported mappings, given for all F in $\mathcal{D}(X, \mathcal{G})$ by

$$\tilde{A}(F): x \to A(x)(F(x)), \qquad x \in X. \tag{1.10}$$

$\tilde{A}$ is then a continuous sum of the operators $A(x)$, $x \in X$. In fact $\tilde{A}$ is an operator from $\mathcal{D}(X, \mathcal{G})$ into itself. More precisely we have

Lemma 1. Let A be in $C^\infty(X, \mathcal{L}(\mathcal{G}))$; for all U in $\mathcal{D}(X, \mathcal{G})$, and for all α in $\mathbb{N}^n$, on the local chart (O_p, φ_p) one has

$$\partial^\alpha(\tilde{A}(U) \circ \varphi_p^{-1})(x) = \sum_{0 \leq \beta \leq \alpha} \binom{\alpha}{\beta} \partial^\beta(A \circ \varphi_p^{-1})(x)(\partial^{\alpha-\beta}(U \circ \varphi_p^{-1})(x))$$

In particular $\tilde{A}(U)$ belongs to $\mathcal{D}(X, \mathcal{G})$.

PROOF. It suffices to prove the assertion when X is an open subset of $\mathbb{R}^n$, that is to say when $X \equiv O_p$, $\varphi_p = \text{id}$. In this case, for y in $X - \{0\}$

and for all real t such that $x + ty$ is in X, one has

$$\lim_{t\to 0} \frac{1}{t} [\tilde{A}(U)(x + ty) - \tilde{A}(U)(x)] = \lim_{t\to 0} \left\{ \frac{1}{t} [A(x + ty) - A(x)][U(x + ty)] + A(x)\left[\frac{1}{t} (U(x + ty) - U(x)) \right] \right\}$$

A being a smooth function on X, $\lim_{t\to 0} (1/t)[A(x + ty) - A(x)]$ exists and is the partial derivative $\partial(A)(x)_y$ of A in x along the vector y; in the same manner the limit $\lim_{t\to 0} (1/t)[U(x + ty) - U(x)]$ exists and is the partial derivative $\partial U(x)_y$ of U in x along y. It follows that

$$\lim_{t\to 0} \frac{1}{t} [\tilde{A}(U)(x + ty) - \tilde{A}(U)(x)] = \partial A(x)_y(U(x)) + A(x)(\partial U(x)_y)$$

In particular, for all $i = 1, 2, \ldots, n$,

$$\frac{\partial}{\partial x_i} [\tilde{A}(U)] = \widetilde{\frac{\partial A}{\partial x_i}} (U) + \tilde{A}\left(\frac{\partial U}{\partial x_i}\right) \tag{1.11}$$

Now using (1.11), the Leibniz rule of derivation, and a recursive argument, one easily deduces that, for any α in $\mathbb{N}^n$, the derivative $\partial^\alpha[\tilde{A}(U)]$ exists and, more precisely, that

$$\partial^\alpha[\tilde{A}(U)] = \sum_{0\le\beta\le\alpha} \binom{\alpha}{\beta} \widetilde{\partial^\beta A}(\partial^{\alpha-\beta} U) \tag{1.12}$$

In particular, $\tilde{A}(U)$ is C^∞ and with compact support, and then $\tilde{A}$ is a linear operator on $\mathcal{D}(X, \mathcal{G})$. □

The natural question arising now concerns the continuity of $\tilde{A}$ with respect to the Schwartz topology of $\mathcal{D}(X, \mathcal{G})$.

Lemma 2. Let A be in $C^\infty(X, \mathcal{L}(\mathcal{G}))$. Given any increasing sequence $k = (k_i)_i$ of positive integers such that $\lim_{i\to+\infty} k_i = +\infty$, there exists a sequence $(M(A)_i)_i$ of strictly positive real numbers such that, for all i in $\mathbb{N}$ and for all U in $\mathcal{D}(X, \mathcal{G})$:

$$|\tilde{A}(U)|_{k_i} \le M(A)_i \cdot |U|_{k_i,i}$$

In particular $\tilde{A}$ is continuous.

PROOF. It suffices to prove the lemma for the case where X is an open subset of $\mathbb{R}^n$. Let $\| \ \|$ be the norm on $\mathcal{L}(\mathcal{G})$ given by

$$\|\lambda\| = \sup_{\substack{u\in\mathcal{G} \\ |u|\le 1}} |\lambda(u)|.$$

It follows from Lemma 1 that for all α in $\mathbb{N}^n$, all U in $\mathcal{D}(X, \mathcal{G})$ and all x in the open subset X of $\mathbb{R}^n$:

$$|\partial^\alpha[\tilde{A}(U)(x)]| \leq \sum_{0 \leq \beta \leq \alpha} \binom{\alpha}{\beta} \|\partial^\beta A(x)\| \cdot |\partial^{\alpha-\beta} U(x)| \tag{*}$$

For each integer i, let

$$q_i^\alpha = \sum_{0 \leq \beta \leq \alpha} \binom{\alpha}{\beta} \cdot \sup_{x \in K_i} \|\partial^\beta \mathrm{A}(\mathrm{x})\|$$

so that $0 \leq q_i^\alpha < +\infty$. Let $M(A)_i$ be the number

$$M(A)_i = \max_{\alpha \in A(n,k_i)} q_i^\alpha$$

Then

$$\max_{\alpha \in A(n,k_i)} \sup_{x \in K_i} |\partial^\alpha[\tilde{A}(U)(x)]| \leq M(A)_i \max_{\beta \in A(n,k_i)} \sup_{x \in K_i} |\partial^\beta U(x)|$$

and

$$|\tilde{A}(U)|_{k_i,i} \leq M(A)_i \cdot |U|_{k_i,i}$$ □

The following Sobolev-type estimate holds:

Lemma 3. Let A be in $C^\infty(X, \mathcal{L}(\mathcal{G}))$, A being constant outside a compact subset of X. For all positive integers k there exists a real number $N_k(A) > 0$ such that, for all U in $\mathcal{D}(X, \mathcal{G})$,

$$\|\tilde{A}(U)\|_2^{(k)} \leq N_k(A) \cdot \|U\|_2^{(k)}$$

In particular $\tilde{A}$ can be extended as a continuous operator on $W_2^{(k)}(X, \mathcal{G})$.

PROOF. Once again, it suffices to prove the assertion when X is an open subset of $\mathbb{R}^n$. Let α be in $A(n, k)$, and for all x in X let $n_\alpha(x)$ be the number $\|\partial^\alpha A(x)\|$, where $\| \ \|$ is the norm on $\mathcal{L}(\mathcal{G})$. Because of the hypothesis on A one has $0 \leq n_\alpha = \sup_{x \in X} n_\alpha(x) < +\infty$. Now, by Lemma 1,

$$|\partial^\alpha[\tilde{A}(U)(x)]| \leq \sum_{0 \leq \beta \leq \alpha} \binom{\alpha}{\beta} |\partial^\beta A(x)(\partial^{\alpha-\beta} U(x))|$$

$$\leq \sum_{0 \leq \beta \leq \alpha} \binom{\alpha}{\beta} n_\beta |\partial^{\alpha-\beta} U(x)|$$

and then

$$\|\partial^\alpha[\tilde{A}(U)]\|_2 \le \sum_{0\le\beta\le\alpha} \binom{\alpha}{\beta} n_\beta \|\partial^{\alpha-\beta}U\|_2$$

$$\le \sum_{0\le\beta\le\alpha} \binom{\alpha}{\beta} n_\beta \|U\|_2^{(k)}$$

It follows that

$$\|\tilde{A}(U)\|_2^{(k)} \le \left[\sum_{\alpha\in A(n,k)} \left\{ \sum_{0\le\beta\le\alpha} n_\beta \binom{\alpha}{\beta} \right\}^2 \cdot (\|U\|_2^{(k)})^2 \right]^{1/2}$$

This assertion is then proved by taking

$$N_k(A) = \left[\sum_{\alpha\in A(n,k)} \left\{ \sum_{0\le\beta\le\alpha} n_\beta \binom{\alpha}{\beta} \right\}^2 \right]^{1/2}. \qquad \square$$

1.5 THE HAUSDORFF SERIES ON $\mathcal{D}(X, \mathcal{G})$

Let exp be the exponential mapping from the Lie algebra $\mathcal{G}$ into the Lie group G. We can associate with $\mathcal{G}$ the formal power series H, so-called Hausdorff series, defined for u, v in $\mathcal{G}$ by

$$H(u, v) = u + v + \tfrac{1}{2}[u, v] + \tfrac{1}{12}[u, [u, v]] + \cdots$$

(see, e.g., [6, Chapter 11, Section 6]).

The following lemma summarizes some properties of H (see [6, Chapter 11, Sections 6–8 and 7, Chapter X]):

Lemma 4. There exist two real numbers q, r with $0 < q < r < (1/3)$ Log (3/2) and such that

1. The restriction $\exp_0$ of exp to $B_r = \{u \in \mathcal{G} / |u| < r\}$ is a diffeomorphism from B_r onto $N_r = \exp(B_r)$, which is a connected open neighborhood of e in G.
2. For all u, v in B_r (resp. in $B_q = \{u \in \mathcal{G} / |u| < q\}$), $[u, v]$ is in B_r (resp. in B_q).
3. There exists an entire function h on $B_q \times B_q$ with values in B_r such that, for all (u, v) in $B_q \times B_q$,

   $$h(u, v) = \exp_0^{-1}(\exp u \cdot \exp v)$$

4. The Hausdorff series $H(U, V)$ with variables U, V is the formal power series coming from h. □

From [8, Chapter 11, Section 6] it follows that, for all pairs (U, V) of elements in $C_0^\infty(X, B_q)$, one has

$$h \circ (U, V) = \sum_{p \geq 1} \eta_p(U, V)$$

in such a manner that, for each integer $p \geq 1$, $\eta_p(U, V)$ is the sum of a number $\alpha_p \leq p - 1$ of terms of the form $\alpha'_s \cdot \eta_p^s(U, V)$, $1 \leq s \leq p - 1$, with $\alpha'_s \leq 1/p$, and with $\eta_p^s(U, V)$ consisting of a multibracket containing s times U and $p - s$ times V.

Lemma 5. Let $\| \ \|$ be an admissible seminorm $\| \ \|$ on $\mathcal{D}(X, \mathcal{G}) = C_0^\infty(X, \mathcal{G})$, i.e., such that, for all U, V in $\mathcal{D}(X, \mathcal{G})$,

$$\|[U, V]\| \leq \|U\| \cdot \|V\|$$

Then for all U, V in $C_0^\infty(X, B_q)$ satisfying $\|U\| \leq \frac{1}{4}$, $\|V\| \leq \frac{1}{4}$ we have $\|h \circ (U \times V)\| \leq 2(\|U\| + \|V\|)$.

PROOF. Let U, V in $C_0^\infty(X, B_q)$; from the above discussion it follows that, for each $p \geq 1$ and $1 \leq s \leq p - 1$:

$$\|\eta_p^s(U, V)\| \leq \|U\|^s \cdot \|V\|^{p-s}$$

and then $\|\eta_p(U, V)\| \leq (\|U\| + \|V\|)^p$. Moreover

$$\|h \circ (U \times V)\| \leq \sum_{p \geq 1} (\|U\| + \|V\|)^p \leq 2(\|U\| + \|V\|)$$

as soon as $\|U\| + \|V\| \leq \frac{1}{2}$. □

Lemma 6. Let $(k_i)_i$ be an increasing sequence of positive numbers with $\lim_{i \to +\infty} k_i = +\infty$. For all U, V in $C_0^\infty(X, B_q)$ with $|U|_{k_i,i} \leq \frac{1}{4}$ and $|V|_{k_i,i} \leq \frac{1}{4}$, one has

$$|h \circ (U \times V)|_{k_i,i} \leq 2(|U|_{k_i,i} + |V|_{k_i,i}).$$

PROOF. From Proposition 1, without loss of generality, we may suppose that, for all U, V in $C_0^\infty(X, \mathcal{G})$, $|[U, V]|_{k_i,i} \leq |U|_{k_i,i} \cdot |V|_{k_i,i}$; the assertion follows then from Lemma 5. □

Lemma 7. Let $k > (1/2)\dim(X)$. For all U, V in $C_0^\infty(X, B_q)$, with $\|U\|_2^{(k)} \leq \frac{1}{4}$ and $\|V\|_2^{(k)} \leq \frac{1}{4}$, one has

$$\|h \circ (U \times V)\|_2^{(k)} \leq 2(\|U\|_2^{(k)} + \|V\|_2^{(k)}).$$

PROOF. Here again, from Proposition 3, we may suppose without loss of generality that, as $k > (1/2)\dim(X)$, $\|[U, V]\|_2^{(k)} \leq \|U\|_2^{(k)} \cdot \|V\|_2^{(k)}$ for all U, V in $\mathcal{D}(X, \mathcal{G})$. The assertion then is a direct consequence of Lemma 5.

1.6 THE NUCLEAR GROUP $\mathcal{D}(X, G)$

Let g be a mapping from the manifold X into the Lie group G whose unit element is denoted e; by the support of g we mean the set $[g]$ consisting of the closure in X of the set $\{x \in X | g(x) \neq e\}$. In particular, if U is a mapping from X into $\mathcal{G}$, one easily sees that U and $\exp U$: $x \to \exp(U(x))$ have the same support.

Let $\mathcal{D}(X, G)$ be the set of all C^∞ – mappings g: $X \to G$ which have a compact support, and for all compact subset K of X, let $\mathcal{D}_K(X, G) = \{g \in \mathcal{D}(X, G) | [g] \subseteq K\}$. One easily sees that $\mathcal{D}(X, G)$ is a group for the pointwise product coming from the product in G, for which the mapping $\varepsilon : x \to \varepsilon(x) = e$, $\forall\, x \in X$ is the unit, and in which $\mathcal{D}_K(X, G)$ is an invariant subgroup.

Let $(K_i)_i$ be the family of compact subsets of X satisfying (1.2); one has

$$\mathcal{D}(X, G) = \bigcup_i \mathcal{D}_{K_i}(X, G).$$

Let us denote by Exp the mapping from $\mathcal{D}(X, \mathcal{G})$ into $\mathcal{D}(X, G)$ given by $\operatorname{Exp} U = \exp \circ U$, $U \in \mathcal{D}(X, \mathcal{G})$; it is clear that, for any compact subsets K of X, Exp maps $\mathcal{D}_K(X, \mathcal{G})$ into $\mathcal{D}_K(X, G)$. Moreover, from Lemma 4, it follows that the restriction of Exp to $C_0^\infty(X, B_q)$ is a bijective mapping from $C_0^\infty(X, B_q)$ onto $C_0^\infty(X, \exp(B_q))$. In order to endow $\mathcal{D}(X, G)$ with a topology which would be analogous to the Schwartz topology of classical test functions spaces, this topology should satisfy two conditions:

1. With this topology $\mathcal{D}(X, G)$ should be a separated topological group.
2. There should exist a neighborhood of the unit element e of $\mathcal{D}(X, G)$ homeomorphic to a neighborhood of O of the nuclear space $\mathcal{D}(X, \mathcal{G})$. In other words $\mathcal{D}(X, G)$ should be a nuclear Lie group in the sense of [1, Chapter 4, Section 5-4].

Let $\mathcal{E}$ be the set of decreasing sequences $\varepsilon = (\varepsilon_i)_i$ of strictly positive numbers such that $\varepsilon_i \leq \frac{1}{4}$ for all i and with $\lim_{i \to +\infty} \varepsilon_i = 0$, and let $\mathcal{K}$ be the set of increasing sequences $k = (k_i)_i$ such that $\lim_{i \to +\infty} k_i = +\infty$; for all (ε, k) in $\mathcal{E} \times \mathcal{K}$, let

$$\mathcal{V}(k, \varepsilon) = \left\{ U \in \mathcal{D}(X, \mathcal{G}) \,\middle|\, \sup_{\alpha \in A(n, k_i)} |u|_i^\alpha \leq \varepsilon_i, \ \forall\, i \in \mathbb{N} \right\}$$

As it is well known (see [2, Chapter 3]), the family $(\mathcal{V}(k, \varepsilon))_{(k,\varepsilon)\in\mathcal{K}\times\mathcal{E}}$ is a fundamental system of neighborhoods of O in $\mathcal{D}(X, \mathcal{G})$. We shall denote by

$$\mathcal{V}_q(k, \varepsilon) \text{ the set } \mathcal{V}(k, \varepsilon) \cap C_0^\infty(X, B_q) \tag{1.13}$$

and by

$$\Gamma(k, \varepsilon) \text{ the set } \mathrm{Exp}(\mathcal{V}_q(k, \varepsilon)) \tag{1.14}$$

for all (k, ε) in $\mathcal{K} \times \mathcal{E}$.

All the $\Gamma(k, \varepsilon)$ contain the unit element e of $\mathcal{D}(X, G)$; the $\mathcal{V}_q(k, \varepsilon)$ are the elements of a fundamental system of neighborhoods of O in the nuclear space $\mathcal{D}(X, \mathcal{G})$.

Lemma 8. $\bigcap_{(k,\varepsilon)\in\mathcal{K}\times\mathcal{E}} \Gamma(k, \varepsilon) = \{e\}$.

PROOF. $g \in \bigcap_{(k,\varepsilon)\in\mathcal{K}\times\mathcal{E}} \Gamma(k, \varepsilon)$ iff there exists U in $\bigcap_{(k,\varepsilon)\in\mathcal{K}\times\mathcal{E}} \mathcal{V}_q(k, \varepsilon)$ such that $g = \mathrm{Exp}\ U$; the assertion follows from the fact that $\bigcap_{(k,\varepsilon)\in\mathcal{K}\times\mathcal{E}} \mathcal{V}_q(k, \varepsilon)$ is reduced to O. □

Lemma 9. Let (k', ε') and (k'', ε'') be in $\mathcal{K} \times \mathcal{E}$; there exists (k, ε) in $\mathcal{K} \times \mathcal{E}$ such that $\Gamma(k, \varepsilon) \subseteq \Gamma(k', \varepsilon') \cap \Gamma(k'', \varepsilon'')$

PROOF. The assertion follows from the fact that we can find k, ε such that $\mathcal{V}_q(k, \varepsilon) \subseteq \mathcal{V}_q(k', \varepsilon') \cap \mathcal{V}_q(k'', \varepsilon'')$. □

Lemma 10. Let (k, ε) be in $\mathcal{K} \times \mathcal{E}$; we can find δ in $\mathcal{E}$ such that $\Gamma(k, \delta) \cdot \Gamma(k, \delta)^{-1} \subseteq \Gamma(k, \varepsilon)$.

PROOF. The $\Gamma(k, \varepsilon)$ are symmetric parts of $\mathcal{D}(X, G)$; we have only to prove that there exists δ in $\mathcal{E}$ such that, for any pair (U, V) in $\mathcal{V}_q(k, \delta) \times \mathcal{V}_q(k, \delta)$, $h \circ (U \times V)$ is in $\mathcal{V}_q(k, \varepsilon)$. From Lemma 6 we can choose $\delta = (\delta_i)_i$ with $\delta_i = \varepsilon_i/4$. □

Lemma 11. Let g be in $\mathcal{D}(X, G)$ and let (k, ε) be in $\mathcal{K} \times \mathcal{E}$; then there exists δ in $\mathcal{E}$ such that $g \cdot \Gamma(k, \delta) \cdot g^{-1} \subseteq \Gamma(k, \varepsilon)$.

PROOF. Here again we can suppose that X is an open subset of $\mathbb{R}^n$. Let $A(g)$ be the continuous sum of the operators $\mathrm{Ad}\, g(x)$, $x \in X$, where Ad denotes the adjoint representation of G into $\mathcal{G}$. An element of $g \cdot \Gamma(k, \delta) \cdot g^{-1}$, for all δ, is of the form $g \cdot \mathrm{Exp}\ U \cdot g^{-1}$, $U \in \mathcal{V}_q(k, \delta)$, i.e. of the form $\mathrm{Exp}[\tilde{A}(g)(U)]$. We have then to prove that there exists δ in $\mathcal{E}$ such that, for all U in $\mathcal{V}_q(k, \delta)$, $\tilde{A}(g)(U)$ is in $\mathcal{V}_q(k, \varepsilon)$. From Lemma 2, we can choose $\delta_i = \varepsilon_i/M(A(g))_i$, $i \in \mathbb{N}$. □

We come now to the main result of this section.

Proposition 4.

1. The family $(\Gamma(k, \varepsilon))_{(k,\varepsilon)\in\mathcal{K}\times\mathcal{E}}$ is a fundamental system of neighborhoods of e for which $\mathcal{D}(X, G)$ is a separated topological group which is a nuclear group

2. The nuclear group $\mathcal{D}(X, G)$ is complete.
3. Let K be a compact subset of X:
   a. The topology of $\mathcal{D}_K(X, G)$ induced by the topology of $\mathcal{D}(X, G)$ is the topology of the C^∞-uniform convergence.
   b. $\mathcal{D}_K(X, G)$ is a closed normal subgroup of $\mathcal{D}(X, G)$.

PROOF. Statement 1 is a trivial consequence of Lemmas 8–11; in particular, Exp is a continuous mapping from $\mathcal{D}(X, \mathcal{G})$ into $\mathcal{D}(X, G)$ for the corresponding Schwartz topologies, and the restriction of Exp to $\mathcal{V}_q(k, \varepsilon)$ is homeomorphism onto $\Gamma(k, \varepsilon)$ for all (k, ε) in $\mathcal{K} \times \mathcal{E}$.

(2) From Lemma 10, $\Gamma(k, \varepsilon)$ being given, there exists $\Gamma(k, \delta)$ such that, for all (U, V) in $\mathcal{V}_q(k, \delta) \times \mathcal{V}_q(k, \delta)$, one has $\text{Exp}\, U \cdot (\text{Exp}\, V)^{-1}$ in $\Gamma(k, \varepsilon)$. It follows that the restriction of Exp to $C_0^\infty(X, B_q)$ is uniformily continuous with respect to the right uniform structure; a similar argument shows that this is also true for the left uniform structure. As the nuclear space $\mathcal{D}(X, \mathcal{G})$ is complete, it follows from Proposition 4, Section 3, Chapter III of [9] that $\mathcal{D}(X, G)$ is complete.

(3) Let K be a compact subset of X; the Schwartz topology of $\mathcal{D}(X, \mathcal{G})$ induces on $\mathcal{D}_K(X, \mathcal{G})$ the topology of the C^∞-uniform convergence (see, e.g., [2, Chapter 1, Section 2]). Assertion 3a follows from the fact that the sets

$$\Gamma_K(k, \varepsilon) = \Gamma(k, \varepsilon) \cap \mathcal{D}_K(X, G) = \text{Exp}(\mathcal{V}_q(k, \varepsilon) \cap \mathcal{D}_K(X, \mathcal{G}))$$

form a fundamental system of neighborhoods of e in $\mathcal{D}_K(X, G)$, for which $\mathcal{D}_K(X, G)$ is complete, hence closed in $\mathcal{D}(X, G)$. Let h be in $\mathcal{D}_K(X, G)$; it is clear that, for all g in $\mathcal{D}(X, G)$, the support of $g \cdot h \cdot g^{-1}$ is contained in the support of h; it follows, then, that $\mathcal{D}_K(X, G)$ is a closed normal subgroup of $\mathcal{D}(X, G)$. □

We call the topology given in (1) of Proposition 4 the *Schwartz topology* for $\mathcal{D}(X, G)$.

1.7 THE LIE ALGEBRA $C_0^k(X, \mathcal{G})$; THE GROUPS $C_0^k(X, G)$

Let k be a nonnegative integer, and let $C_0^k(X, \mathcal{G})$ be the set of all $\mathcal{G}$-valued compactly supported functions of class C^k on X. Endowed with the pointwise algebraic structure coming from that of $\mathcal{G}$, $C_0^k(X, \mathcal{G})$ is a Lie algebra. In the same way, the set $C_0^k(X, G)$ of all the G-valued compactly supported functions of class C^k on X, endowed with the pointwise product coming from that of G, is a group.

Let K be a compact subset of X; one easily sees that $C_K^k(X, G) = \{g \in C_0^k(X, G)/[g] \subseteq K\}$ is a normal subgroup of $C_0^k(X, G)$; in the same way the space $C_K^k(X, \mathcal{G}) = \{U \in C_0^k(X, \mathcal{G})/[U] \subseteq K\}$ is an ideal of $C_0^k(X, \mathcal{G})$.

For $\varepsilon = (\varepsilon_i)_i$ in $\mathscr{E}$, let $\mathscr{V}^k(\varepsilon) = \{U \in C_0^k(X, \mathscr{G}) / |U|_{k,i} \leq \varepsilon_i, \forall\, i \in \mathbb{N}\}$. The proof in Section 1.2 shows that the $\mathscr{V}^k(\varepsilon)$, $\varepsilon \in \mathscr{E}$, form a fundamental system of neighborhoods of O for which $C_0^k(X, \mathscr{G})$ is a Fréchet space, inductive limit of the Banach spaces $C_{K_i}^k(X, \mathscr{G})$, the Lie algebra structure being compatible with this topology.

Now let us consider the sets

$$\Gamma^k(\varepsilon) = \operatorname{Exp}(C_0^k(X, B_q) \cap \mathscr{V}^k(\varepsilon)), \qquad \varepsilon \in \mathscr{E}.$$

It follows from Section 1.6 that the $\Gamma^k(\varepsilon)$ constitute a fundamental system of neighborhoods of e in $C_0^k(X, G)$ for which $C_0^k(X, G)$ becomes a separated topological group. We have

$$\begin{aligned} C_0^k(X, G) &= \bigcup_{i \in \mathbb{N}} C_{K_i}^k(X, G) \\ \mathscr{D}(X, G) &= \bigcap_{k \in \mathbb{N}} C_0^k(X, G) \end{aligned} \tag{1.15}$$

1.8 THE HILBERT-SOBOLEV LIE GROUPS $W_2^{(k)}(X, G)$

Let k be an integer such that $k > (1/2)\dim(X)$, and for any ε such that $0 < \varepsilon < \frac{1}{4}$ let us consider the sets

$$\mathscr{V}_k(\varepsilon) = \{U \in C_0^\infty(X, B_q) \mid \|U\|_2^{(k)} < \varepsilon\}$$

and

$$\Gamma_k(\varepsilon) = \operatorname{Exp}(\mathscr{V}_k(\varepsilon)).$$

The $\mathscr{V}_k(\varepsilon)$, $\frac{1}{4} \geq \varepsilon > 0$, form a basis for the neighborhoods of O in $\mathscr{D}(X, \mathscr{G})$ with respect to the pre-Hilbertian structure given by $\|\ \|_2^{(k)}$ (cf. Section 1.3). It follows that the sets:

$$\overline{\mathscr{V}}_k(\varepsilon) = \{U \in W_2^{(k)}(X, \mathscr{G}) \mid U(X) \subseteq B_q, \|U\|_2^{(k)} < \varepsilon\}, \qquad \varepsilon > 0$$

form a basis for the neighborhoods of O in $W_2^{(k)}(X, \mathscr{G})$ with respect to its structure of Hilbert-Sobolev Lie algebra (cf. Section 1.3). From the group point of view one has:

Proposition 5. Let $k \in \mathbb{N}$ such that $k > (1/2)\dim(X)$. $\{\Gamma_k(\varepsilon), \frac{1}{4} \geq \varepsilon > 0\}$ is a fundamental system of neighborhoods of e in $\mathscr{D}(X, G)$ for which it is a separated topological group which is connected if G is connected. Moreover $\operatorname{Exp}\colon \mathscr{D}(X, \mathscr{G}) \to \mathscr{D}(X, G)$ is continuous with respect to the topology given by $\|\ \|_2^{(k)}$.

PROOF. Without loss of generality we can suppose that X is an open subset of $\mathbb{R}^n$ endowed with the Lebesgue measure dx. As we have

$\bigcap_{\varepsilon>0} \mathcal{V}_k(\varepsilon) = (O)$ it follows that

$$\bigcap_{\varepsilon>0} \Gamma_k(\varepsilon) = \bigcap_{\varepsilon>0} \operatorname{Exp}(\mathcal{V}_k(\varepsilon)) = (\operatorname{Exp} O) = (e).$$

Let $\varepsilon', \varepsilon''$ be real numbers with $0 < \min(\varepsilon', \varepsilon'') \leq \max(\varepsilon', \varepsilon'') \leq \frac{1}{4}$ and let $\varepsilon = \min(\varepsilon', \varepsilon'')$; one easily sees that $\Gamma_k(\varepsilon) \subseteq \Gamma_k(\varepsilon') \cap \Gamma_k(\varepsilon'')$.

Let $\frac{1}{4} \geq \varepsilon > 0$; let us prove that there exists $\delta > 0$, with $\delta \leq \frac{1}{4}$ such that $\Gamma_k(\delta) \cdot \Gamma_k(\delta)^{-1} \subseteq \Gamma_k(\varepsilon)$. As the Γ_k are symmetric we need only to prove that $\Gamma_k(\delta) \cdot \Gamma_k(\delta) \subseteq \Gamma_k(\varepsilon)$ or, equivalently, that we can find δ, $0 < \delta \leq \frac{1}{4}$, such that $\|U\|_2^{(k)} < \delta$ and $\|V\|_2^{(k)} < \delta$ imply $\|h \circ (U \times V)\|_2^{(k)} < \varepsilon$. From Lemma 7 it follows that this is possible by taking $\delta = \varepsilon/4$.

Let g be in $\mathfrak{D}(X, G)$ and let $\frac{1}{4} \geq \varepsilon > 0$; from Lemma 3 there exists $N_k > 0$ such that $\|\widetilde{A(g)}(U)\|_2^{(k)} \leq N_k \cdot \|U\|_2^{(k)}$. Now, taking $\delta = \min(\frac{1}{4}, \varepsilon/N_k)$, it follows that, for all Exp U in $\Gamma_k(\delta)$, $g^{-1} \cdot \operatorname{Exp} U \cdot g = \operatorname{Exp}(\widetilde{A(g)}U)$ is in $\Gamma_k(\varepsilon)$.

We have then proved that the $\Gamma_k(\varepsilon)$ are a fundamental system of neighborhoods of e in $\mathfrak{D}(X, G)$ for which $\mathfrak{D}(X, G)$ is a separated topological group, and for which Exp is continuous with respect to the metrizable topology given by $\| \ \|_2^{(k)}$. If G is connected, $\operatorname{Exp}(\mathfrak{D}(X, \mathcal{G}))$ spans $\mathfrak{D}(X, G)$; Exp being continuous and $\mathfrak{D}(X, \mathcal{G})$ being connected, it follows from Proposition 7, Chapter III, Section 2 of [9], that $\mathfrak{D}(X, G)$ is connected. □

For the topology given by the $\Gamma_k(\varepsilon)$, $\frac{1}{4} \geq \varepsilon > 0$, $\mathfrak{D}(X, G)$ is a metrizable group. Let us denote by $W_2^{(k)}(X, G)$ the complete group spanned by $\mathfrak{D}(X, G)$ with respect to this topology. We shall denote by $\overline{\Gamma_k(\varepsilon)}$ the set $\operatorname{Exp}(\overline{\mathcal{V}}_q(\varepsilon))$, $\frac{1}{4} \geq \varepsilon > 0$.

Proposition 6. Let k be an integer such that $k > (1/2)\dim(X)$. $W_2^{(k)}(X, G)$ is a Hilbert Lie group with the Hilbert-Sobolev Lie algebra $W_2^{(k)}(X, \mathcal{G})$ as $W_2^{(k)}(X, G)$ is the completion of the current group $\mathfrak{D}(X, G)$ with respect to the Sobolev metric given by $\| \ \|_2^{(k)}$.

PROOF. The last part in the proof of Proposition 5 allows us to assert the uniform continuity of the mapping $g \to g^{-1}$ in $C_0^\infty(X, \exp(B_q))$; it follows from Proposition 9, Section 3, Chapter III of [9] that $W_2^{(k)}(X, G)$ has the $\overline{\Gamma_k}(\varepsilon)$, $\varepsilon > 0$, as fundamental system of neighborhoods of e. Moreover, as Exp: $\mathfrak{D}(X, \mathcal{G}) \to \mathfrak{D}(X, G)$ is continuous with respect to the Sobolev metric given by $\| \ \|_2^{(k)}$, its extension to $W_2^{(k)}(X, \mathcal{G})$ with values in $W_2^{(k)}(X, G)$ is continuous.

As $k > (1/2)\dim(X)$, from [5], $W_2^{(k)}(X, G)$ endowed with the Sobolev metric given by $\| \ \|_2^{(k)}$ has a structure of Hilbert manifold such that, for f in $W_2^{(k)}(X, G)$, the tangent space in f to $W_2^{(k)}(X, G)$ is the space of mappings φ from X into the tangent bundle TG of G such that $\varphi(x)$ is in $T_{f(x)}G$ and such that the generalized derivatives of order $\leq k$ are square dx-integrable. In particular, the tangent space at e is exactly the Hilbert-Sobolev Lie algebra $W_2^{(k)}(X, \mathcal{G})$. From well-known properties of Sobolev manifolds [5], one easily deduces that $(g_1, g_2) \to g_1 \cdot g_2^{-1}$ is a C^∞-mapping from $W_2^{(k)}(X, G) \times W_2^{(k)}(X, G)$ into $W_2^{(k)}(X, G)$, and that $W_2^{(k)}(X, G)$ is a Hilbert Lie group with Lie algebra $W_2^{(k)}(X, \mathcal{G})$. □

$W_2^{(k)}(X, G)$, as given by Proposition 6 is called a *Hilbert-Sobolev Lie group*.

1.9 THE GAUGE GROUPS AND THEIR VARIOUS SOBOLEV COMPLETIONS OF ORDER 1

When the Lie group G is a compact semisimple Lie group, $\mathcal{D}(X, G)$ is called a *gauge group*. The interest for these groups arises in several domains (see, e.g., [10], Section 2); in particular, in mathematical physics these groups appear in connection with Yang-Mills models and their quantization (see, e.g., [11]). From the mathematical point of view their nice properties come from the fact that they act unitarily on the spaces of $\mathcal{G}$-valued 1-forms on X, and from the use of the so-called Maurer-Cartan cocycle.

Let $\Omega(X)$ be the space of real-valued and compactly supported smooth 1-forms on X; $\mathcal{D}_1(X, \mathcal{G}) = \Omega(X) \otimes \mathcal{G}$ is the space of compactly supported smooth sections of the smooth vector bundle Hom $(TX, \mathcal{G})$, or, equivalently, the space of $\mathcal{G}$-valued compactly supported smooth 1-forms on X. Let us consider henceforth $\mathcal{G}$ endowed with its canonical scalar product $\langle\ ,\ \rangle$ given by the opposite of its Killing form, and let us suppose $\Omega(X)$ is endowed with a pre-Hilbertian structure given by a positive definite quadratic form q; q and $\langle\ ,\ \rangle$ give rise in a natural way to a scalar product $\langle\ ,\ \rangle_q$ on $\mathcal{D}_1(X, \mathcal{G}) = \Omega(X) \otimes \mathcal{G}$ which is invariant under the action of $\mathcal{D}(X, G)$ on $\mathcal{D}_1(X, \mathcal{G})$ given by

$$(g, \omega) \to V(g)\omega, \qquad g \in \mathcal{D}(X, G), \quad \omega \in \mathcal{D}_1(X, \mathcal{G}),$$

where $V(g)\omega$ is the 1-form

$$x \to (V(g)\omega)(x) = \mathrm{Ad}\, g(x) \cdot \omega(x), \qquad x \in X \tag{1.16}$$

$V: g \to V(g)$ is then an orthogonal representation of $\mathcal{D}(X, G)$ into $\mathcal{D}_1(X, \mathcal{G})$.

The Maurer-Cartan cocycle $\beta: \mathcal{D}(X, G) \to \mathcal{D}_1(X, \mathcal{G})$ is defined by

$$\beta(g) = dg \cdot g^{-1}, \qquad \text{for all } g \in \mathcal{D}(X, G) \tag{1.17}$$

$dg \cdot g^{-1}$ denoting the right translation by g^{-1} of the element $dg \in \mathfrak{D}(TX)$.

Lemma 12. For all g_1, g_2 in $\mathfrak{D}(X, G)$, one has

$$\beta(g_1 \cdot g_2) = \beta(g_1) + V(g_1)\beta(g_2)$$

PROOF. This important formula can be found e.g. in [12, Chapter XIX, Section 15]. We give here a direct algebraic proof. Let x be in X; from (1.17) and the rules of derivation one has

$$\begin{aligned}
\beta(g_1 \cdot g_2)(x) &= d(g_1 \cdot g_2)(x) \cdot g_2^{-1}(x) \cdot g_1^{-1}(x) \\
&= [(dg_1)_x \cdot g_2(x) + g_1(x) \cdot (dg_2)_x] \cdot g_2^{-1}(x) \cdot g_1^{-1}(x) \\
&= (dg_1)_x \cdot g_1^{-1}(x) + g_1(x) \cdot ((dg_2)_x \cdot g_2^{-1}(x)) \cdot g_1^{-1}(x) \\
&= \beta(g_1)(x) + \operatorname{Ad} g_1(x) \cdot \beta(g_2)(x)
\end{aligned}$$

and then

$$\beta(g_1 \cdot g_2) = \beta(g_1) + V(g_1) \cdot \beta(g_2) \qquad \square$$

Corollary. Let q be a positive definite quadratic form on $\Omega(X)$, invariant under the pointwise adjoint representation given by (1.16), and let $|\cdot|_q = (\langle\, .\, ,\, .\, \rangle_q)^{1/2}$ denote the corresponding Euclidean norm on $\mathfrak{D}_1(X, \mathcal{G})$. Let us suppose that X is a noncompact manifold. The mapping d_q: $\mathfrak{D}(X, G) \times \mathfrak{D}(X, G) \to \mathbb{R}^+$ given by $d_q(g_1, g_2) = |\beta(g_1 \cdot g_2^{-1})|$ is a metric on $\mathfrak{D}(X, G)$ which is invariant by right translations.

PROOF. Let us recall that $\beta(g) = dg \cdot g^{-1}$, and that $|\ |_q$ is a norm. It follows that $|\beta(g)|_q = 0$ iff g is constant on X; as we have supposed that X is noncompact, this implies that $g = e$; then $|\beta(g)|_q = 0$ iff $g = e$. One easily deduces that $d_q(g_1, g_2) = 0$ iff $g_1 \cdot g_2^{-1} = e$, i.e., iff $g_1 = g_2$.

Moreover, from Lemma 12, one gets, for any g in $\mathfrak{D}(X, G)$: $\beta(g^{-1}) = -V(g^{-1})\beta(g)$, and then $|\beta(g^{-1})|_q = |\beta(g)|_q$. It follows that, for all g_1, g_2 in $\mathfrak{D}(X, G)$,

$$\begin{aligned}
d_q(g_2, g_1) &= |\beta(g_2 \cdot g_1^{-1})|_q = |\beta((g_1 \cdot g_2^{-1})^{-1})|_q \\
&= |\beta(g_1 \cdot g_2^{-1})|_q = d_q(g_1, g_2).
\end{aligned}$$

Now let g_1, g_2, g_3 be in $\mathfrak{D}(X, G)$:

$$\begin{aligned}
d_q(g_1, g_3) &= |\beta(g_1 \cdot g_3^{-1})|_q = |\beta(g_1 \cdot g_2^{-1} \cdot g_2 \cdot g_3^{-1})|_q \\
&= |\beta(g_1 \cdot g_2^{-1})_q + V(g_1 \cdot g_2^{-1})\beta(g_2 \cdot g_3^{-1})|_q \\
&\leq |\beta(g_1 \cdot g_2^{-1})|_q + |\beta(g_2 \cdot g_3^{-1})|_q
\end{aligned}$$

because $V(g_1 \cdot g_2^{-1})$ is a unitary operator with respect to $|\ |_q$; it follows that

$d_q(g_1, g_3) \leq d_q(g_1, g_2) + d_q(g_2, g_3)$. Moreover, for all g in $\mathfrak{D}(X, G)$, one has

$$\begin{aligned} d_q(g_1 \cdot g, g_3 \cdot g) &= |\beta(g_1 \cdot g \cdot g^{-1} \cdot g_3^{-1})|_q \\ &= |\beta(g_1 \cdot g_3^{-1})|_q = d_q(g_1, g_3) \end{aligned}$$

Hence, d_q is invariant by right translations. □

REMARK. In the same way, one can prove that the mapping δ_q: $(g_1, g_2) \to \delta_q(g_1, g_2) = |\beta(g_1^{-1} \cdot g_2)|_q$ is a left invariant metric on $\mathfrak{D}(X, G)$.

For all $\varepsilon > 0$, let us denote by B_ε^q the set

$$B_\varepsilon^q = \{g \in \mathfrak{D}(X, G) / |\beta(\gamma g \gamma^{-1})|_q < \varepsilon, \qquad \forall\, \gamma \in \mathfrak{D}(X, G)\}.$$

All the B_ε^q contain e, and then the family $\mathcal{F}^q$ of the B_ε^q, $\varepsilon > 0$, forms a basis for a filter in $\mathfrak{D}(X, G)$ (see, e.g., [9, Chapter 1, Section 6-3]). The problem is now to determine whether $\mathfrak{D}(X, G)$ is a metric group with respect to d_q (or δ_q).

Proposition 7. $\mathfrak{D}(X, G)$ is a metric group with respect to d_q.

PROOF. From [9, Chapter III, Section 1-2], we need to prove that:

(i) For all $\varepsilon > 0$, there exists $\varepsilon' > 0$ such that $B_{\varepsilon'}^q \cdot B_{\varepsilon'}^q \subseteq B_\varepsilon^q$.
(ii) For all $\varepsilon > 0$, there exists $\varepsilon'' > 0$ such that $(B_{\varepsilon''}^q)^{-1} \subseteq B_\varepsilon^q$.
(iii) For all γ in $\mathfrak{D}(X, G)$, and all $\varepsilon > 0$, there exists $\varepsilon_\gamma > 0$ such that $B_{\varepsilon_\gamma}^q \subseteq \gamma \cdot B_\varepsilon^q \cdot \gamma^{-1}$.

(i) Let $\varepsilon > 0$, and let $\varepsilon' = \varepsilon/2$; an element of $B_{\varepsilon'}^q \cdot B_{\varepsilon'}^q$ is of form $g_1 \cdot g_2$ with

$$|\beta(\gamma g_1 \gamma^{-1})|_q < \varepsilon/2 \qquad \text{for all } \gamma \text{ in } \mathfrak{D}(X, G)$$

and

$$|\beta(\gamma g_2 \gamma^{-1})|_q < \varepsilon/2 \qquad \text{for all } \gamma \text{ in } \mathfrak{D}(X, G)$$

It follows that, for all γ in $\mathfrak{D}(X, G)$,

$$\begin{aligned} |\beta(\gamma g_1 g_2 \gamma^{-1})|_q &= |\beta(\gamma g_1 \gamma^{-1} \cdot \gamma g_2 \gamma^{-1})|_q \\ &= |\beta(\gamma g_1 \gamma^{-1}) + V(\gamma g_1 \gamma^{-1})\beta(\gamma g_2 \gamma^{-1})|_q \\ &\leq |\beta(\gamma g_1 \gamma^{-1})|_q + |\beta(\gamma g_2 \gamma^{-1})|_q \\ &< \varepsilon/2 + \varepsilon/2 \end{aligned}$$

which shows that $B_{\varepsilon/2}^q \cdot B_{\varepsilon/2}^q \subseteq B_\varepsilon^q$ for all $\varepsilon > 0$.

(ii) In the proof of the corollary of Lemma 12 we have seen that for all g' in $\mathfrak{D}(X, G)$ one has $\beta(g'^{-1}) = -V(g'^{-1})\beta(g')$ from which it follows that $|\beta(g'^{-1})|_q = |\beta(g')|_q$, and then $(B_\varepsilon^q)^{-1} = B_\varepsilon^q$.

(iii) Let $\varepsilon > 0$ and let γ be in $\mathcal{D}(X, G)$; we have to prove that we can find ε_γ such that $\gamma^{-1} \cdot B^q_{\varepsilon_\gamma} \cdot \gamma \subseteq B^q_\varepsilon$. Let g be in $\gamma^{-1} \cdot B^q_\varepsilon \cdot \gamma$: g is of the form $g = \gamma^{-1} \cdot g' \cdot \gamma$ with $|\beta(\gamma' g' \gamma'^{-1})|_q < \varepsilon$ for all γ' in $\mathcal{D}(X, G)$; in particular, $|\beta(\gamma' \gamma g (\gamma' \gamma)^{-1})|_q < \varepsilon$ for all γ', from which one easily deduces that g is in B^q_ε; so (iii) is proved by taking $\varepsilon_\gamma = \varepsilon$. □

REMARK. $(\mathcal{D}(X, G), d_q)$ is a separated topological group.

Proposition 8. The completion $H^1_q(X, G)$ of $\mathcal{D}(X, G)$ with respect to d_q is a complete metric group.

PROOF. In order to prove that $H^1_q(X, G)$ is a complete metric group, from [9, Chapter III, Section 3, Proposition 9], it suffices to prove that the mapping $\varphi: g \to g^{-1}$ is uniformily continuous on B^q_ε, $\varepsilon > 0$. For any g, g' such that $g \cdot g'^{-1} \in B^q_\varepsilon$, $d_q(g, g') = |\beta(g \cdot g'^{-1})|_q < \varepsilon$ and $d_q(\varphi(g), \varphi(g')) = |\beta(g^{-1} \cdot g')|$; but $g \cdot g'^{-1} \in B^q_\varepsilon$ implies $|\beta(g^{-1} \cdot g \cdot g'^{-1} \cdot g)|_q = |\beta(g'^{-1} \cdot g)|_q < \varepsilon$. It follows that $d_q(\varphi(g'), \varphi(g)) < \varepsilon$, as soon as $g \cdot g'^{-1}$ is in B^q_ε. This completes the proof. □

REMARK. The Hilbert-Sobolev Lie groups $W_2^{(k)}(X, G)$ exist for any finite-dimensional Lie group G, compact semisimple or not, if $k > (1/2)\dim(X)$ (cf. Section 1.8), and these completions of $\mathcal{D}(X, G)$ depend only on the selected volume measure dx on X.

When G is compact semisimple, the Sobolev q-completion $H^1_q(X, G)$ of $\mathcal{D}(X, G)$ gives another way, in the case $k = 1$ without the above condition on $\dim(X)$, to get by completion of $\mathcal{D}(X, G)$ a complete metric group, the metric depending on the first derivative of the G-valued test functions. This construction has been given in the particular case where q comes from a Riemannian structure on X in [13] and [14]. It is, however, not canonical inasmuch as $H^1_q(X, G)$ depends on q which does not depend necessarily only on the volume measure on X, as we shall see in the chapter devoted to the energy representations of gauge groups. Moreover, we cannot assert that $H^1_q(X, G)$ is a Hilbert Lie group if $\dim(X) > 1$.

1.10 ON SOME SUBGROUPS OF $\mathcal{D}$(X, G) AND RELATED TOPICS

Let Γ be either G or $\mathcal{G}$, and let k be a nonnegative integer; for all elements u of $\mathcal{D}(X, \Gamma)$ and x in X, we shall denote by $j^k_x(u)$ the k-jet of u with source x. The set $J^k(X, \Gamma)$ of all k-jets of elements of $\mathcal{D}(X, \Gamma)$ has a canonical structure of C^∞-bundle manifold with base X [15, Chapter XVI, Section 5]. For x in X and γ in Γ, $J^k_x(X, \Gamma)_\gamma$ denotes the submanifold of k-jets with source x and target γ;

$$J^k_x(X, \Gamma) = \bigcup_{\gamma \in \Gamma} J^k_x(X, \Gamma)_\gamma$$

is the fiber above x. One also knows (see, e.g., [15]) that for any u in $\mathcal{D}(X, \Gamma)$ the mapping $j^k u: x \to j_x^k(u)$ is a compactly supported C^∞-section of the bundle manifold $J^k(X, \Gamma)$.

The manifolds $J_x^k(X, G)$ and $J_x^k(X, \mathcal{G})$ have been studied quite extensively (see, e.g., [16] and [17, Chapter 1]). We summarize here the main results:

(i) $J_x^k(X, G)$ is a real and finite-dimensional Lie group, the product being given by

$$j_x^k(g) \cdot j_x^k(g') = j_x^k(g \cdot g'),$$

the unit element being $j_x^k(e)$.

(ii) $J_x^k(X, \mathcal{G})$ is the Lie algebra of $J_x^k(X, G)$, the bracket being given by

$$[j_x^k(U), j_x^k(V)] = j_x^k([U, V]).$$

(iii) $J_x^k(X, G)_e$ is a normal nilpotent subgroup of $J_x^k(X, G)$.

(iv) For x in X and g in $\mathcal{D}(X, G)$, let us denote by g_x the C^∞-mapping from X into G given by

$$g_x(y) = g(y) \cdot g(x)^{-1}.$$

The mapping $j_x^k(g) \to (g(x), j_x^k(g_x))$ is an isomorphism of Lie groups from $J_x^k(X, G)$ onto the semidirect product $G \cdot J_x^k(X, G)_e$. In particular, $J_x^1(X, G)$ is isomorphic to $G \times \mathcal{G}^{\dim(X)}$.

For any integer $k \geq 0$ and each x in X, let us denote by $G^k(x)$ the set $\{g \in \mathcal{D}(X, G) | j_x^k(g) = j_x^k(e)\}$.

Lemma 13. $G^k(x)$ is a normal subgroup of $\mathcal{D}(X, G)$ and $J_x^k(X, G)$ is isomorphic to $\mathcal{D}(X, G)/G^k(x)$.

PROOF. Let g_1, g_2 be in $G^k(x)$;

$$\begin{aligned} j_x^k(g_1 \cdot g_2^{-1}) &= j_x^k(g_1) \cdot j_x^k(g_2^{-1}) = j_x^k(g_1) \cdot (j_x^k(g_2))^{-1} \\ &= j_x^k(e) \cdot j_x^k(e) = j_x^k(e) \end{aligned}$$

and then $g_1 \cdot g_2^{-1}$ is in $G^k(x)$; moreover for all g in $G^k(x)$ and all γ in $\mathcal{D}(X, G)$

$$j_x^k(\gamma \cdot g \cdot \gamma^{-1}) = j_x^k(\gamma) \cdot j_x^k(e) \cdot (j_x^k(\gamma))^{-1} = j_x^k(\gamma) \cdot (j_x^k(\gamma))^{-1} = j_x^k(e)$$

and then $\gamma \cdot g \cdot \gamma^{-1}$ belongs to $G^k(x)$.

Now let g_1, g_2 be in $\mathcal{D}(X, G)$; one has $g_1 \equiv g_2 \bmod G^k(x)$ iff $g_1 \cdot g_2^{-1}$ belongs to $G^k(x)$, that is to say iff $j_x^k(g_1) = j_x^k(g_2)$; this completes the proof. □

Let us consider now the group Aut(G) of all automorphisms of G, and the group Diff(X) of all the diffeomorphisms of X. The direct product group Aut (G) $\times$ Diff(X) acts on $\mathfrak{D}(X, G)$ such that, for (a, φ) in Aut (G) $\times$ Diff(X), for all g in $\mathfrak{D}(X, G)$, $(a, \varphi) \cdot g$ is the element of $\mathfrak{D}(X, G)$ given by

$$((a, \varphi) \cdot g)(x) = a[(g \circ \varphi)(x)], \qquad x \in X.$$

For a given (a, φ) in Aut (G) $\times$ Diff(X), let us denote by $\mathfrak{D}_{(a,\varphi)}(X, G)$ the set

$$\mathcal{G}_{(a,\varphi)}(X, G) = \{g \in \mathfrak{D}(X, G)/(a, \varphi) \cdot g = g\}$$

One easily sees that

Lemma 14. $\mathfrak{D}_{(a,\varphi)}(X, G)$ is a subgroup of $\mathfrak{D}(X, G)$.

REMARK. In the case where X is the one-dimensional torus T, a is a non-inner automorphism of G of order p, $p \geq 2$, and φ a diffeomorphism of T of order p, the group $\mathfrak{D}_{(a,\varphi)}(T, G)$ is closely connected with the so-called Kac-Moody groups; see Chapter 5.

REFERENCES

1. I. M. Gelfand and N. Y. Vilenkin, *Les Distributions*, vol. 4, Dunod, Paris (1967), *Generalized Functions*, vol. 4, Academic Press, New York (1964).
2. L. Schwartz, *Théorie des Distributions*, Hermann, Paris (1966).
3. F. Treves, *Topological Vector Spaces, Distribution and Kernels*, Academic Press, New York (1967).
4. L. S. Sobolev, *Applications of Functional Analysis in Mathematical Physics*, Translations of Mathematical Monographs, vol. 7, Amer. Math. Soc., Providence, RI (1963).
5. R. S. Palais, *Foundations of Global Nonlinear Analysis*, Benjamin, New York (1968).
6. N. Bourbaki, *Groupes et Algèbres de Lie*, Chapters II–III, Hermann, Paris (1972).
7. G. Hochschild, *The Structure of Lie Groups*, Holden Day, San Francisco (1965); *La structure des Groupes de Lie*, Dunod, Paris (1968).
8. N. Bourbaki, *Variétés Différentielles et Analytiques*, Sections 8–15, Hermann, Paris (1971).
9. N. Bourbaki, *Topologie Générale*, Chapters I–IV, Hermann, Paris (1971).
10. S.Albeverio and R. Høegh-Krohn, *Functional Analysis and Markov Processes* (M. Fukushima, ed.), Lecture Notes in Math., vol. 923, Springer-Verlag, Berlin, Heidelberg, New York, pp. 133–145 (1982).
11. I. Segal, *J. Functional Anal. 33*: 175–194 (1979).
12. J. Dieudonné, *Eléments d'Analyse*, vol. 4, Gauthier-Villars, Paris (1971), *Treatise on Analysis*, vol. 4, Academic Press, New York (1974)
13. S. Albeverio and R. Høegh-Krohn, *Compositio Math.*, *36*:37–52 (1978).

14. A. M. Vershik, I. M. Gelfand, and M. I. Graev, *Dokl. Akad. Nauk.*, *232*:745–748 (1977). English transl., *Soviet Math. Dokl.*, *18*:118–121 (1977). *Compositio Math.*, *35*:299–334 (1977); *42*:217–243 (1981).
15. J. Dieudonné, *Eléments d'Analyse*, vol. 3, Gauthier-Villars, Paris (1970); *Treatise on Analysis*, vol. 3, Academic Press, New York (1972).
16. K. R. Parthasarathy and K. Schmidt, *Comm. Math. Phys.*, *50*:167–175 (1976).
17. J. Marion, Sur les représentations unitaires d'ordre *k* des groupes $C_0^\infty(X, G)$, preprint, Marseille-Luminy (1978).

2

Multiplicative G-Distributions on a Riemannian Manifold

2.1 INTRODUCTION

Let X be a Riemannian manifold and let G be a real Lie group. The nuclear group $\mathcal{D}(X, G)$ may be viewed as a noncommutative generalization of a test function space like $\mathcal{D}(X, \mathbb{R})$. Let us consider a real distribution T on X; the mapping $e^{i\langle T,\cdot\rangle}$ is a continuous unitary character on $\mathcal{D}(X, \mathbb{R})$; so, following an idea of Gelfand, one may view the classes of irreducible continuous unitary representations of the nuclear group $\mathcal{D}(X, G)$ as (noncommutative) (multiplicative) distributions on X.

This chapter is devoted to the definition of a rigorous setting for a noncommutative extension of the concept of multiplicative distribution on a manifold. The basic concepts of order and support and the basic tools, in particular the so-called representations of type (S), are investigated. The first problem to be solved is how to get irreducible unitary representations of $\mathcal{D}(X, G)$ with infinite support. In this chapter, partial answers in the case G nilpotent and G semisimple noncompact are given.

2.2 BASIC MATERIALS FOR A THEORY OF G-DISTRIBUTIONS ON A RIEMANNIAN MANIFOLD

2.2.1 The Concept of G-Distribution

Let X be a Riemannian manifold endowed with a volume measure dx, and let G be a real Lie group with Lie algebra $\mathcal{G}$.

Definition 1. A *continuous unitary representation* of the nuclear Lie group $\mathcal{D}(X, G)$ into the Hilbert space $\mathcal{H}$ is a homomorphism π from $\mathcal{D}(X, G)$ into the unitary group of $\mathcal{H}$, such that for all u, v in $\mathcal{H}$, the mapping $g \to \langle \pi(g)u, v \rangle$ is continuous on $\mathcal{D}(X, G)$ (where $\langle \, , \, \rangle$ is here the scalar product in $\mathcal{H}$).

We shall denote by $\hat{\mathcal{D}}(X, G)$ the set of classes of unitary equivalence of all the irreducible continuous unitary representations of $\mathcal{D}(X, G)$.

Definition 2. The elements of $\hat{\mathcal{D}}(X, G)$ are called *the G-distributions* on the manifold X.

We can extend the concept of continuous representation of $\mathcal{D}(X, G)$ to a Banach setting in the following way.

Definition 3. A *continuous representation* of $\mathcal{D}(X, G)$ into a Banach space $\mathcal{B}$ is a homomorphism π from $\mathcal{D}(X, G)$ into the group of continuous automorphisms on $\mathcal{B}$ such that, for all Φ in the dual space $\mathcal{B}'$ of $\mathcal{B}$ and for all u in $\mathcal{B}$, $g \to \Phi(\pi(g)u)$ is continuous on $\mathcal{D}(X, G)$.

2.2.2 The $\mathbb{R}$-Distributions

As we shall see later, the study of G-distributions is a difficult problem, and, a fortiori, the complete knowledge of $\hat{\mathcal{D}}(X, G)$ for a general Lie group G is far from being achieved. The abelian case $G = \mathbb{R}$ is discussed in [10, Chapter 4, Section 5]. See also, e.g., [28]. We shall call the elements of $\hat{\mathcal{D}}(X, \mathbb{R})$ $\mathbb{R}$-distributions on X. To determine them, let us consider an irreducible and continuous unitary representation π of $\mathcal{D}(X, \mathbb{R})$, where $\mathbb{R}$ denotes the field of real numbers, into some Hilbert space $\mathcal{H}$ with scalar product $(\, , \,)$, and let h be a cyclic vector for π. The mapping $\varphi: g \to (\pi(g)h, h)$, $g \in \mathcal{D}(X, \mathbb{R})$ is a continuous function of positive type on $\mathcal{D}(X, \mathbb{R})$ and we have $\varphi(0) = (h, h) = 1$; from the Bochner-Minlos theorem (see, e.g., [10, Chapter 4, Section 4, Theorem 1], [16]), it follows that there exists a normed measure μ on the space of distributions $\mathcal{D}'(X, \mathbb{R})$ (i.e., the space of real-valued distributions) such that its Fourier transform $\hat{\mu}$ is given by

$$\hat{\mu}(g) = \varphi(g) = \int_{\mathcal{D}'(X,\mathbb{R})} e^{i\langle T, g \rangle} \, d\mu(T)$$

Now, using the spectral decomposition of the unitary operators $\pi(g)$, one easily concludes that $\mathcal{H}$ can be realized as the space $L^2(\mathcal{D}'(X, \mathbb{R}); \mu)$, and the operators $\pi(g)$ as operators of multiplication by $e^{i\langle T, g \rangle}$ for some T in $\mathcal{D}'(X, \mathbb{R})$. As π is irreducible, one concludes that π is a unitary character of $\mathcal{D}(X, \mathbb{R})$, of the form $\chi_T: g \to e^{i\langle T, g \rangle}$ for some distribution T in $\mathcal{D}'(X, \mathbb{R})$. It follows that the following holds:

Proposition 1. The $\mathbb{R}$-distributions on the manifold X are the continuous unitary characters of the additive group $\mathcal{D}(X, \mathbb{R})$; they are of the form χ_T: $g \rightarrow \exp(i \langle T, g \rangle)$, where T is any distribution on X which belongs to $\mathcal{D}'(X, \mathbb{R})$.

The bijective mapping χ: $\mathcal{D}'(X, \mathbb{R}) \rightarrow \hat{\mathcal{D}}(X, \mathbb{R})$, given by $\chi(T) = \chi_T$ allows to identify each real distribution T with its multiplicative form χ_T. Taking into account the fact that the objects which generalize the continuous unitary characters of an abelian group are the classes of unitary equivalence of continuous unitary representations of a general group, we can hope that the G-distributions on the manifold X are good candidates to give a generalization of the multiplicative real distributions on X.

We have first to extend to the G-distributions the usual concepts of support and order of the theory of classical distributions (see, e.g., [23]).

2.2.3 Support of a G-Distribution

Definition 4. Let π be a continuous representation of $\mathcal{D}(X, G)$ into a Banach space $\mathcal{B}$; an element x of X is said to be in the *support* of π if, for any open neighborhood V of x, there exists g in $\mathcal{D}(X, G)$ with support contained in V and such that g does not belong to $\ker(\pi)$.

We remark that the support of the unitary character χ_T of $\mathcal{D}(X, \mathbb{R})$, with T in $\mathcal{D}'(X, \mathbb{R})$, is equal to the support of the distribution T in the usual sense.

The case of the $\mathbb{R}$-distributions was solved quite early and stated as such in 1961 by Gelfand and Vilenkin (see [10, Chapter 4, Section 5]). For the case of $\mathcal{D}(X, G)$, where G is not an abelian group, it was only in 1973 that the first irreducible and continuous unitary representation of $\mathcal{D}(X, SL(2, \mathbb{R}))$ with support the whole manifold X was given (see [25] and [26]).[a] The representations with finite support are called since then *located representations*; the representations with support the whole manifold X are called *nonlocated representations* (see also the discussion in [9]).

Lemma 1. Any two unitarily equivalent unitary representations of $\mathcal{D}(X, G)$ have the same support.

PROOF. Let π and π' be two unitarily equivalent unitary representations of $\mathcal{D}(X, G)$, and let A be a unitary operator intertwining π and π'. An element x of X is in the support of π' iff for any open neighborhood V of x we can find g in $\mathcal{D}(X, G)$ with support in V and such that $\pi'(g) \neq 1$. But $\pi'(g) \neq 1$ iff $\pi(g) = A^{-1} \cdot \pi'(g) \cdot A$ is not the identity. It follows that x is in the support of π' iff x is in the support of π.

[a] A basic paper in the noncommutative case is [9].

The above lemma allows us to define the support of a G-distribution.

Definition 5. The *support* $[\omega]$ of a G-distribution ω on X is the common support of the continuous irreducible unitary representations of $\mathcal{D}(X, G)$ which are in class ω.

Example 1. Let ω be a class of unitary equivalence of continuous irreducible unitary representations of the Lie group G, and let π be an element of the class ω. To each element x of X we can associate a continuous and irreducible unitary representation π_x of $\mathcal{D}(X, G)$, given for all g in $\mathcal{D}(X, G)$ by

$$\pi_x(g) = \pi(g(x)).$$

Obviously, if π' is unitarily equivalent to π, π'_x is unitarily equivalent to π_x; it follows that the class of π_x is a G-distribution which depends only on the class ω of π; we denote this G-distribution by ω_x; of course the support of ω_x reduces to $\{x\}$. More generally, a finite subset $\{x_1, x_2, \ldots, x_p\}$ of X being given, $\pi_{x_1} \otimes \pi_{x_2} \otimes \cdots \otimes \pi_{x_p}$ is a continuous irreducible unitary representation of $\mathcal{D}(X, G)$, and its class is a G-distribution $\omega_{x_1,x_2,\ldots,x_p}$ with support $\{x_1, x_2, \ldots, x_p\}$.

2.2.4 Order of a G-Distribution

Let k be a nonnegative integer; a continuous representation θ of the Fréchet Lie group $C_0^k(X, G)$ (see Chapter 1, Section 7) restricted to its dense subgroup $\mathcal{D}(X, G)$ yields a continuous representation of $\mathcal{D}(X, G)$ because the Schwartz topology is stronger than the topology on $\mathcal{D}(X, G)$ induced by the topology of $C_0^k(X, G)$; moreover, owing to the density of $\mathcal{D}(X, G)$ in $C_0^k(X, G)$, if θ is irreducible, its restriction to $\mathcal{D}(X, G)$ remains irreducible.

Let us consider two continuous irreducible unitary representations π and π' arising by restriction from two continuous irreducible unitary representations θ and θ' of respectively $C_0^k(X, G)$ and $C_0^{k'}(X, G)$, k and k' being the smallest integers having this property, and let us suppose that π and π' are unitarily equivalent. Owing to their continuity and the fact that $\mathcal{D}(X, G)$ is dense both in $C_0^k(X, G)$ and $C_0^{k'}(X, G)$, it follows that θ and θ' are unitarily equivalent and that $k = k'$. This allows us to introduce, as for the usual distributions, the concept of order of a continuous representation and of a G-distribution.

Definition 5. (a) A continuous representation π of $\mathcal{D}(X, G)$ is said to be of *finite order* if there exists an integer $k \geq 0$ such that π is the restriction of some continuous representation of $C_0^k(X, G)$, and the smallest integer for which the above property remains true is called the *order* of π.

(b) A G-distribution ω on X is said to be of *order* k if k is the order of one (and then of all) of the irreducible continuous unitary representations of $\mathcal{D}(X, G)$ which are in the class of ω.

Example 2. In the case of the $\mathbb{R}$-distributions χ_T, T in $\mathcal{D}'(X, \mathbb{R})$, it is clear that χ_T is of order k if and only if the distribution T is of order k in the sense of usual distributions.

Example 3. Let k be some positive integer, and let x_0 be in X; we endow the finite-dimensional Lie group of k-jets $J_{x_0}^k(X, G)$ (see Chapter 1, Section 10) with the left-invariant Haar measure dj. Let us consider the left regular representation L^k of $J_{x_0}^k(X, G)$ into $L^2(J_{x_0}^k(X, G); dj)$; we get a continuous unitary representation of $\mathcal{D}(X, G)$, denoted $L_{x_0}^k$, into $L^2(J_{x_0}^k(X, G); dj)$, given by

$$L_{x_0}^k(g) = L^k(j_{x_0}^k(g)), \qquad g \in \mathcal{D}(X, G)$$

It is clear that $L_{x_0}^k$ is a nonirreducible continuous unitary representation of $\mathcal{D}(X, G)$ of order k and with support $\{x_0\}$.

From a theoretical point of view it is not difficult to get a G-distribution of order $\leq k$ and with finite support: let $\{x_1, x_2, \ldots, x_p\}$ be a finite subset in X, and, for each $i = 1, 2, \ldots, p$, let π_i be a continuous and irreducible unitary representation of the k-jets group $J_{x_i}^k(X, G)$; we get a continuous and irreducible unitary representation $\pi_{x_i}^k$ of $\mathcal{D}(X, G)$ such that, for all g in $\mathcal{D}(X, G)$,

$$\pi_{x_i}^k(g) = \pi_i(j_{x_i}^k(g)), \qquad i = 1, 2, \ldots, p$$

which is of order $\leq k$ and with support $\{x_i\}$, and one easily deduces that the class of unitary equivalence of $\pi_{x_1}^k \otimes \pi_{x_2}^k \otimes \cdots \otimes \pi_{x_p}^k$ is a G-distribution on X of order $\leq k$ and support $\{x_1, x_2, \ldots, x_p\}$. It follows that the problem of finding G-distributions of finite order and with finite support is reduced to the problem of finding the continuous irreducible unitary representations of the k-jets groups $J_x^k(X, G)$, $k \geq 0$, $x \in X$. The first work about such representations was done in 1967 by Gelfand and Graev in the case $G = SU(2)$ [9].

Let us recall that $J_x^k(X, G)$ is a regular semidirect product of G by the nilpotent Lie group $J_x^k(X, G)_e$ (Chapter 1, Section 10, [4]); it follows that, theoretically, Mackey's method (see, e.g., [19]) gives a way in order to get all the continuous and irreducible unitary representations of $J_x^k(X, G)$; however, for $k \geq 2$, computations are very complicated [1].

2.2.5 Multiplicative G-Integrals

In the theory of usual distributions the measures are the distributions of order zero; if μ is a real measure on X, the $\mathbb{R}$-distribution $\chi_\mu : g \to \exp(i\mu(g))$ is a multiplicative integral on X and χ_μ is of order zero in the sense of $\mathbb{R}$-distributions. This leads to the following generalization.

Definition 7. The G-distributions on X of order zero are called *G-multiplicative integrals.*

Examples 4. For all real-valued measures μ on X, the corresponding $\mathbb{R}$-distribution χ_μ is a multiplicative $\mathbb{R}$-integral on X.

Example 5. Let ω be a class of irreducible and continuous unitary representations of G, and let x_0 be in X; the G-distribution ω_{x0} with support $\{x_0\}$ (see Example 1) is a multiplicative G-integral on X; we can consider ω_{x0} as a multiplicative generalization of the Dirac measure centered in x_0.

Example 6. Let G be the "$at + b$ group," that is, the group of affine transformations of the straight line leaving invariant its orientation. The elements of $\mathfrak{D}(X, G)$ can be realized as pairs, (A, B), with B in $\mathfrak{D}(X, \mathbb{R})$ and with A in the set of C^∞-mappings from X into $\mathbb{R}$ such that $A(x) > 0$ for all x in X and $A(x) = 1$ outside a compact subset of X, the product law $\cdot$ being given by

$$(A, B) \cdot (A', B') = (AA', AB' + B)$$

For any strictly positive measure μ on X:

$$\theta_\mu(A, B) \equiv \exp\{i \int_X \ln(A(x))\, d\mu(x)\}$$

is a multiplicative G-integral on X, whose support coincides with the support of μ.

2.3 THE ACTION OF Diff(X) ON $\hat{\mathfrak{D}}$(X, G)

2.3.1 Description of the Action

The group Diff(X) of all diffeomorphisms of X acts on $\mathfrak{D}(X, G)$ by

$$(\varphi, g) \to \varphi \cdot g = g \circ \varphi, \qquad g \in \mathfrak{D}(X, G), \quad \varphi \in \mathrm{Diff}(X)$$

It follows that a continuous representation π of $\mathfrak{D}(X, G)$ being given, each element Ψ of Diff(X) yields a continuous representation $\Psi \cdot \pi$ of $\mathfrak{D}(X, G)$ such that for any g in $\mathfrak{D}(X, G)$

$$(\Psi \cdot \pi)(g) = \pi(g \circ \Psi)$$

Moreover, if π is a unitary representation, then $\Psi \cdot \pi$ is also unitary.

Lemma 2. Let π be a continuous unitary representation of $\mathcal{D}(X, G)$ into some Hilbert space $\mathcal{H}$, and let Ψ be in Diff(X); then

(i) $\Psi \cdot \pi$ is irreducible if and only if π is irreducible;

(ii) π' being another continuous unitary representation of $\mathcal{D}(X, G)$, $\Psi \cdot \pi$ and $\Psi \cdot \pi'$ are unitarily equivalent if and only if π and π' are unitarily equivalent.

PROOF. (i) Let $\mathcal{L}$ be a closed subspace of $\mathcal{H}$ invariant under π. For all u in $\mathcal{L}$, $\pi(g \circ \Psi)(u)$ is in $\mathcal{L}$, and then $\mathcal{L}$ is invariant by $\Psi \cdot \pi$; the converse remains true owing to the fact that $g \to g \circ \Psi$ is an automorphism of the group $\mathcal{D}(X, G)$. This then proves (*i*).

(ii) Let $\mathcal{H}'$ be the Hilbert space of the unitary representation π'. π and π' are unitarily equivalent if and only if there exists an isometric operator A from $\mathcal{H}$ onto $\mathcal{H}'$ such that, for all g' in $\mathcal{D}(X, G)$, $A\pi(g') = \pi'(g')A$. This holds if and only if, for all g in $\mathcal{D}(X, G)$, one has $A\pi(g \circ \Psi) = \pi'(g \circ \Psi)A$, which in turn holds if and only if $\Psi \cdot \pi$ and $\Psi \cdot \pi'$ are unitarily equivalent. Moreover, one sees that π and π', as well as $\Psi \cdot \pi$ and $\Psi \cdot \pi'$, have the same intertwining operators.

For a continuous irreducible unitary representation π of $\mathcal{D}(X, G)$, let us denote by $\hat{\pi}$ the corresponding G-distribution; from Lemma 2 it follows that the next corollary holds:

Corollary. Diff(X) acts on $\hat{\mathcal{D}}(X, G)$ by $(\Psi, \hat{\pi}) \to \Psi \cdot \hat{\pi} = \widehat{\Psi \cdot \pi}$.

2.3.2 The Groups Diff(X; ω)

The group Diff(X) acts on the space $\mathcal{D}'(X, \mathbb{R})$ of real distributions on X by $(\Psi, T) \to \Psi \cdot T$, $\Psi \in$ Diff(X), $T \in \mathcal{D}'(X, \mathbb{R})$ such that, for all f in $\mathcal{D}(X, \mathbb{R})$

$$\langle \Psi \cdot T, f \rangle = \langle T, f \circ \Psi \rangle$$

For a given distribution T in $\mathcal{D}'(X, \mathbb{R})$, we can consider the group

$$\text{Diff}(X; T) = \{\Psi \in \text{Diff}(X) | \Psi \cdot T = T\}$$

Conversely, in order to know T, it would suffice to know Diff(X; T). In the same way, to each G-distribution ω, we can associate the group

$$\text{Diff}(X; \omega) = \{\Psi \in \text{Diff}(X) | \Psi \cdot \omega = \omega\}$$

Definition 8. Let ω be a G-distribution on X, and let T be a real distribution on X; we shall say that ω is a *multiplicative distribution* with respect to T if Diff(X; T) $\subseteq$ Diff(X; ω).

Example 7. Let ω be a class of unitary equivalence of continuous irreducible unitary representations of G, and let x_0 be in X. Let δ_{x_0} be the Dirac measure at the point x_0, and let ω_{x_0} be multiplicative G-integral arising from ω (see Example 5).
One easily sees that

$$\mathrm{Diff}(X; \delta_{x_0}) = \mathrm{Diff}(X; \omega_{x_0}) = \{\Psi \in \mathrm{Diff}(X) | \Psi(x_0) = x_0\}.$$

It follows that ω_{x_0} is a multiplicative distribution (or order zero and support $\{x_0\}$) with respect to the Dirac measure δ_{x_0}.

More details about the connection between the G-distributions ω and their corresponding group $\mathrm{Diff}(X; \omega)$ can be found in [20].

Gelfand and Guichardet started the problem of the multiplicative integral (see, e.g., [13]) which is at the present time still only partially solved, and can be formulated as follows: a positive measure μ with support X and a Lie group G being given, what are the multiplicative G-integrals which are multiplicative integrals with respect to μ? This problem is a particular case of the following one: an element T of $\mathcal{D}'(X, \mathbb{R})$ with support X being given, what are the multiplicative G-distributions with respect to T?

2.4 A BASIC TOOL: REPRESENTATIONS OF TYPE (S)

At the present time, all known nonlocated irreducible unitary representations of the groups $\mathcal{D}(X, G)$ can be realized as representations "of type (S)," which are closely connected with Fock spaces and the 1-cohomology of $\mathcal{D}(X, G)$ (see, e.g., [11, Section 3.2]). To describe them we need first to explain the concepts of symmetric Hilbert spaces and 1-cohomology of groups.

2.4.1 Symmetric Hilbert spaces

Let $\mathcal{H}$ be an Hilbert space over the field $\mathbb{K} = \mathbb{R}$ or $\mathbb{C}$ with scalar product $\langle\ ,\ \rangle$. For all integers $n \geq 0$ let us denote by $S^n\mathcal{H}$ the following Hilbert space: if $n = 0$, $S^0\mathcal{H} = \mathbb{K}$; if $n \geq 1$, $S^n\mathcal{H}$ denotes the closed subspace of $\mathcal{H}^{\otimes n}$ spanned by the elements of the form $h^{\otimes n} = h \otimes h \otimes \cdots \otimes h$, $h \in \mathcal{H}$, $S^n\mathcal{H}$ is then the nth symmetric tensor power of $\mathcal{H}$.

By definition the *symmetric Hilbert space*, or *symmetric Fock space associated with* $\mathcal{H}$, is the Hilbert space $S\mathcal{H} = \oplus_{n\geq 0} S^n\mathcal{H}$. An element h of $\mathcal{H}$ being given, we shall denote by EXP h the following element of $S\mathcal{H}$:

$$\begin{aligned}\mathrm{EXP}\, h &= 1 \oplus h \oplus (2!)^{-1/2}\, h^{\otimes 2} \oplus \cdots \oplus (n!)^{-1/2}\, h^{\otimes n} \oplus \cdots \\ &= \sum_{n\geq 0}^{\oplus} (n!)^{-1/2}\, h^{\otimes n}\end{aligned}$$

We shall denote by EXP $\mathcal{H}$ the subset of $S\mathcal{H}$ such that

$$\text{EXP } \mathcal{H} = \{\text{EXP } h | h \in \mathcal{H}\}$$

We denote by (,) the natural scalar product in EXP $\mathcal{H}$, defined by extension from (EXP h; EXP h') $= \exp\{\langle h, h'\rangle\}$. The following proposition summarizes the basic properties of symmetric Hilbert spaces.

Proposition 2. Let $\mathcal{H}$ be an Hilbert space.

(i) EXP $\mathcal{H}$ is a total set in $S\mathcal{H}$, and consists of linearly independent vectors.
(ii) The mapping $h \to$ EXP h is one-to-one and bicontinuous from $\mathcal{H}$ into $S\mathcal{H}$.
(iii) Let $\mathcal{H}$ be the direct sum of two Hilbert spaces: $\mathcal{H} = \mathcal{H}_1 \oplus \mathcal{H}_2$; the mapping $\text{EXP}(h_1 \oplus h_2) \to \text{EXP } h_1 \otimes \text{EXP } h_2$ is extended into a canonical isomorphism from $S(\mathcal{H}_1 \oplus \mathcal{H}_2)$ onto $S\mathcal{H}_1 \otimes S\mathcal{H}_2$.

PROOF. (i), (ii), and (iii) are proved in [11, Chapter 2, Propositions 2.1, 2.2, and 2.3 respectively].

In a very important case $S\mathcal{H}$ can be realized as an L^2-space with respect to a Gaussian measure.

Proposition 3. Let E be a separable real nuclear space which is the strict inductive limit of Fréchet spaces, endowed with a continuous positive definite quadratic form θ. Let $\mathcal{H}$ be the complex Hilbert space spanned by E with respect to θ, let E' be the dual space of E, and let μ be the Gaussian measure on E' defined by its Fourier transform:

$$\hat{\mu}: F \to \hat{\mu}(F) = \int_{E'} e^{i\langle \chi, F\rangle}\, d\mu(\chi) = \exp\left(-\frac{1}{2}\,\theta\,(F)\right)$$

The mapping EXP $h \to \exp\{\theta\,(h)/2 + i\,\langle\cdot, h\rangle\}$, $h \in \mathcal{H}$, can be extended into an isomorphism of Hilbert spaces from $S\mathcal{H}$ onto $L^2(E'; \mu)$.

PROOF. The proof of this classical result of Segal can be found for instance in [11, Chapter 7, Theorem 7-1].

2.4.2 1-Cohomology for Topological Groups

Definition 9. Let Γ be a topological group, and let π be a continuous representation of Γ into a Banach space $\mathbb{B}$. A continuous mapping b: $\Gamma \to \mathbb{B}$ is called a *1-cocycle* of Γ with respect to π if, for all γ and γ' in Γ:

$$b(\gamma \cdot \gamma') = b(\gamma) + \pi(\gamma) \cdot b(\gamma')$$

The vector space of continuous 1-cocycles of Γ with respect to π is denoted by $Z^1(\Gamma, \pi)$; the subspace of 1-coboundaries, that is to say the subspace consisting on mappings of the form $\gamma \to \pi(\gamma)v - v$, where $v \in \mathbb{B}$, is denoted by $B^1(\Gamma, \pi)$, and the coset $H^1(\Gamma, \pi) = Z^1(\Gamma, \pi)|B^1(\Gamma, \pi)$ is called the 1-cohomology group of Γ with respect to π.

An important question is the following: a topological group Γ being given, do there exist continuous unitary representations of Γ such that $H^1(\Gamma, \pi) \neq (0)$? How can such representations be constructed? In the case where $\Gamma = G$ is a finite-dimensional Lie group, an important number of papers are devoted to the study of this question (see, e.g., [6,12,17,27,28]). It turns out that the solution of this problem is closely connected with the question whether G has the Kazhdan property: Γ has the *Kazhdan property* if the trivial representation of Γ is an isolated point in the dual ([8]) $\hat{\Gamma}$ of Γ. If $\Gamma = G$ is a Lie group we have the following result.

Proposition 4. $H^1(G, \pi) = (0)$ for all unitary representations (irreducible or not) if and only if G has the Kazhdan property.

PROOF. The fact that the Kazhdan property implies $H^1(G, \pi) = (0)$ for all unitary representations π of G was proved in [17, Theorem V-1]; the converse was proved in [6].

Our knowledge of the cohomology of $\Gamma = \mathfrak{D}(X, G)$ is rather limited: in the case G compact semisimple, one knows that the Maurer-Cartan cocycle (see Chapter 1, Lemma 12) is a 1-cocycle with respect to the representation V of $\mathfrak{D}(X, G)$ in the space of 1-forms $\mathfrak{D}_1(X, \mathfrak{G})$, and we shall see in the next chapter that this 1-cocycle is nontrivial; in [21] we gave another example of nontrivial 1-cocycle with respect to the adjoint representation of $\mathfrak{D}(X, G)$ (see also the next chapter). One knows also various (but unfortunately not unitarizable) representations of $\mathfrak{D}(X, G)$ for which the various derivatives of the Maurer-Cartan cocycle are nontrivial 1-cocycles (see [21,28]).

Let us come back to the case $\Gamma = G$; the following proposition summarizes the principal results about the unitary 1-cohomology of usual Lie groups.

Proposition 5. Let G be a finite-dimensional real Lie group; we assume that G is connected.

1. If G is nilpotent, then the trivial representation has a nontrivial 1-cohomology.
2. If G is simple, $G \neq SO(n, 1)$, and $G \neq SU(n, 1)$, then $H^1(G, \pi) = (0)$ for all irreducible unitary representations π of G.

3. If G is one of the groups $SO(n, 1)$ (resp. $SU(n, 1)$), there exist one (resp. two) irreducible unitary representations of G with nontrivial 1-cohomology.
4. If G is a compact semisimple group, then $H^1(G, \pi) = (0)$ for any unitary representation π of G.
5. If G is a motion group, there exists a finite-dimensional irreducible unitary representation π of G such that $H^1(G, \pi) \neq (0)$.
6. If G is the Poincaré group, there exists an irreducible unitary representation π of G such that $H^1(G, \pi) \neq (0)$.

PROOF. Statements (1), (2), and (3) are proved in [6]; (4) follows from the fact that a continuous 1-cocycle on a compact group is bounded and, then, is necessarily, a 1-coboundary. Statement (5) is proved in [14], and (6) in [12].

2.4.3 Unitary Representations of Type (S)

Let us begin with an important result about the class $U_S(\mathcal{H})$ of unitary operators on the symmetric Hilbert space $S\mathcal{H}$ of a complex Hilbert space $\mathcal{H}$ which preserve the set $\{\lambda \text{ EXP } h, \lambda \in \mathbb{C}, h \in \mathcal{H}\}$, the proof of which was given in [13, Lemma 2.1 and Theorem 2.1].

Lemma 3. Let $\mathcal{H}$ be a complex Hilbert space, let $\mathbb{T}$ be the torus of unimodular complex numbers, and let $\mathcal{U}(\mathcal{H})$ be the unitary group of $\mathcal{H}$. Let $A \in \mathcal{U}(\mathcal{H})$, $c \in \mathbb{T}$, b: $\mathbb{T} \to \mathcal{H}$ a continuous cocycle. To any such triple (c, A, b) we can associate an element $U_{c,A,b}$ of $U_S(\mathcal{H})$ such that, on EXP $\mathcal{H}$:

$$U_{c,A,b}(\text{EXP}\, h) = c\exp\{-\tfrac{1}{2}\|b\|^2 - \langle Ah, b\rangle\}\,\text{EXP}(Ah + b), \qquad h \in \mathcal{H}.$$

If (c', A', b') is another element of $\mathbb{T} \times \mathcal{U}(\mathcal{H}) \times \mathcal{H}$, we have

$$U_{c,A,b} \cdot U_{c',A',b'} = \exp\{i\, \text{Im}\langle b, Ab'\rangle\} U_{cc',AA',b+Ab'}$$

(i.e., $U_{c,A,b}$ is a topological group with $\cdot$ as multiplication and $U_{(1,1,0)}$ as unit element).

From the above lemma we get three corollaries permitting us to construct new unitary representations of a given topological group.

Corollary 1. Let π be a continuous unitary representation of a topological group Γ into a complex Hilbert space $\mathcal{H}$, let b: $\Gamma \to \mathcal{H}$ be a continuous 1-cocycle of Γ with respect to π, and let c: $\Gamma \to \mathbb{T}$ be a continuous mapping such that, for any γ, γ' in Γ:

$$c(\gamma \cdot \gamma') = \exp\{i\, \text{Im}\langle b(\gamma), \pi(\gamma)b(\gamma')\rangle\} \cdot c(\gamma) \cdot c(\gamma')$$

Then $U_{c,\pi,b}$: $\gamma \to U_{c(\gamma),\pi(\gamma),b(\gamma)}$ is a continuous unitary representation of Γ into $S\mathcal{H}$.

Definition 10. The unitary representations of a topological group Γ of the form $U_{c,\pi,b}$ are called *unitary representations of type (S).*

Corollary 2. Let π, b, and Γ be as in Corollary 1, and let us suppose moreover that, for all γ, γ' in Γ, $\mathrm{Im}\langle b(\gamma), b(\gamma')\rangle = 0$ then $U_{1,\pi,b} \equiv \mathrm{EXP}_b\, \pi$: $\gamma \to U_{1,\pi(\gamma),b(\gamma)}$ is a continuous unitary representation of type (S) of Γ.

PROOF. $\mathrm{Im}\,\langle b(\gamma), \Pi(\gamma) b(\gamma')\rangle = \mathrm{Im}\,\langle b(\gamma), b(\gamma\,\gamma') - b(\gamma)\rangle = 0$ owing to the hypothesis, and then $c \equiv 1$ satisfies the hypothesis of Corollary 1.

As a particular case, one has

Corollary 3. Let π be a continuous unitary representation of Γ into a real Hilbert space $\mathcal{H}_0$, and let b be an element of $Z^1(\Gamma, \pi)$; we extend π to a unitary representation, again denoted π, of Γ into the complexified Hilbert space $\mathcal{H}$ of $\mathcal{H}_0$. Then $\mathrm{EXP}_b\pi = U_{1,\pi,b}$ is a unitary representation of type (S) of Γ into $S\mathcal{H}$.

In the case where Γ acts unitarily on some real nuclear space E which is a strict inductive limit of Fréchet spaces, according to some scalar product (,) on E, if we denote by $\mathcal{H}$ the complex Hilbert space spanned by E with respect to (,) and by E' the dual of E, we can realize $S\mathcal{H}$ as the space $L^2(E', \mu)$, where μ is the Gaussian measure on E' with Fourier transform

$$\hat{\mu}: F \to \hat{\mu}(F) = \int_{E'} e^{i\langle \chi, F\rangle}\, d\mu(\chi) = e^{-1/2(F,F)}$$

by Proposition 3.

The unitary action of Γ on E is extended to a unitary representation π of Γ into the real Hilbert space $\mathcal{H}_0$ spanned by E, and then π is extended to a continuous representation, denoted again by π, of Γ into E' such that for all χ in E':

$$\langle \pi(\gamma)\chi, F\rangle = \langle \chi, \pi(\gamma^{-1})F\rangle, \ \gamma \in \Gamma, \ F \in E$$

Proposition 6. With the above conditions, if b is a 1-cocycle of Γ with respect to π, the representation $\mathrm{EXP}_b\pi$ of type (S) of Corollary 3 can be realized in $L^2(E', \mu)$ as follows: for all γ in Γ, Φ in $L^2(E', \mu)$, one has, for all χ in E',

$$(\mathrm{EXP}_b\, \pi(\gamma)\Phi)(\chi) = e^{i\langle \chi, b(\gamma)\rangle}\, \Phi(\pi(\gamma^{-1})\chi)$$

PROOF. From Proposition 3, one has an isomorphism θ of Hilbert spaces from $S\mathcal{H}$ onto $L^2(E', \mu)$ such that on EXP $\mathcal{H}$

$$\theta(\text{EXP } h) = \exp\{\tfrac{1}{2}\|h\|^2 + i\langle ., h\rangle\}.$$

On the other hand, one has, in the $S\mathcal{H}$ picture

$$\text{EXP}_b\ \pi(\gamma)(\text{EXP } h) = \exp\{-\tfrac{1}{2}\|b(\gamma)\|^2 - \langle \pi(\gamma)h, b(\gamma)\rangle\}\ \text{EXP}\,(\pi(\gamma)h + b(\gamma)),$$

and Proposition 6 follows by an easy computation.

Proposition 7. Let π be a continuous unitary representation of Γ into the Hilbert space $\mathcal{H}$, and let b and b' be two elements of $Z^1(\Gamma, \pi)$ satisfying the condition of Corollary 2. If b and b' are in the same class of 1-cohomology, then $\text{EXP}_b\pi$ and $\text{EXP}_{b'}\pi$ are unitarily equivalent.

PROOF. Let b' and b be in the same class of 1-cohomology. There exists v in $\mathcal{H}$ such that, for all γ in Γ,

$$b'(\gamma) = b(\gamma) + \pi(\gamma)\, v - v$$

Let us consider now the operator $\text{EXP}_{-v}I = U_{1,I,-v}$, where I is the identity on $\mathcal{H}$. From Lemma 3 it follows that $\text{EXP}_{-v}I$ is a unitary operator on $S\mathcal{H}$ and that, for all g in Γ,

$$\text{EXP}_{-v}\ I \cdot \text{EXP}_{b(g)}\pi(g) = \text{EXP}_{b(g)-v}\pi(g) = \text{EXP}_{b'(g)-\pi(g)v}\pi(g)$$

and

$$\text{EXP}_{b'(g)}\pi(g) \cdot \text{EXP}_{-v}I = \text{EXP}_{b'(g)-\pi(g)v}\pi(g)$$

It follows that the unitary operator $\text{EXP}_{-v}I$ intertwines $\text{EXP}_b\pi$ and $\text{EXP}_{b'}\pi$. □

From the above proposition, it follows that, for any 1-coboundary b, $\text{EXP}_b\pi$ is unitarily equivalent to $\text{EXP}_0\pi$.

Lemma 4. Let π be a continuous unitary representation of Γ into the Hilbert space $\mathcal{H}$; $\text{EXP}_0\pi = \bigoplus_{n\geq 0} \pi^{\otimes n}$, where $\pi^{\otimes n}$ is the unitary representation of Γ into the nth symmetric tensor power $S^n\mathcal{H}$ such that, for all γ in Γ and h in $\mathcal{H}$, $\pi^{\otimes n}(\gamma)(h^{\otimes n}) = (\pi(\gamma)h)^{\otimes n}$. In particular $\text{EXP}_0\pi$ is always reducible.

PROOF. The assertion follows from the fact that, for all γ_r in Γ and h in $\mathcal{H}$, $\text{EXP}_0\pi(\gamma)(\text{EXP } h) = \text{EXP}\,(\pi(\gamma)h)$, and from the fact that, by definition, for all elements u in $\mathcal{H}$

$$\text{EXP } u = \sum_{n\geq 0}^{\oplus} (n!)^{-1/2} u^{\otimes n}.$$

Corollary. A necessary condition in order to get an irreducible representation $\text{EXP}_b\pi$ of type (S) is that $H^1(\Gamma, \pi) \neq (0)$ and b not be a 1-coboundary.

PROOF. This follows easily from Proposition 6 and Lemma 4.

Let us consider a representation $\text{EXP}_b\pi$ of type (S) of Γ into $S\mathcal{H}$; an easy computation shows that, for all γ in Γ,

$$\text{EXP}_b\pi(\gamma)(\text{EXP } 0) = \exp\{-\tfrac{1}{2}\|b(\gamma)\|^2\} \text{ EXP}(b(\gamma))$$

and then:

$$(\text{EXP}_b\pi(\gamma)(\text{EXP } 0), \text{EXP } 0) = \exp\left\{-\frac{1}{2}\|b(\gamma)\|^2\right\}.$$

Of course, $\varphi_{\pi,b}: \gamma \rightarrow \varphi_{\pi,b}(\gamma) = \exp\{-1/2\ \|b(\gamma)\|^2\}$ is a continuous positive definite function on Γ (see, e.g., [8, Section 13-4-6]), the one associated with the unitary representation $\text{EXP}_b\pi$ and the vacuum vector EXP 0.

Definition 11. The continuous positive definite function $\varphi_{\pi,b}$ on Γ associated with the representation of type (S), $\text{EXP}_b\pi$, and to the vacuum vector is called the *continuous positive definite function of type (S) associated to* $\text{EXP}_b\pi$.

REMARK. A unitary representation $\text{EXP}_b\pi$ of type (S) being given, the vacuum vector EXP 0 is not necessarily a cyclic vector for $\text{EXP}_b\pi$, as it was shown for instance in [11, Remark 4.2, Section 4.2].

2.5 NONLOCATED MULTIPLICATIVE *G*-INTEGRALS; THE CONTINUOUS TENSOR PRODUCTS OF REPRESENTATIONS OF *G*

Let X be endowed with a nonatomic positive measure μ with support the whole manifold X, and let G be a Lie group. The construction of non located multiplicative G-integrals with respect to μ is based on the theory of continuous tensor products of Hilbert spaces (see, e.g., [3, 11, 15]) and on the construction of an analog of infinitely divisible representations (see, e.g., [2, 22, 24]). These representations, which are of type (S), are the so-called continuous tensor products of unitary representations of G (c.t.p.u.r.).

The first c.t.p.u.r. giving a multiplicative G-integral were constructed by Vershik, Gelfand, and Graev in the case $G = SL(2, \mathbb{R})$ ($\cong SU(1, 1)$) [25], and for $G = SO(n; 1)$ or $SU(n; 1)$ in [26]. Later, Delorme [7] and Karpushev [18] gave conditions of irreducibility of the c.t.p.u.r. in a more general setting.

2.5.1 The Construction of c.t.p.u.r

Definition 12. Let X be endowed with a nonatomic positive measure μ with support X, and let G be a Lie group. We shall say that $[(\mathcal{H}_x)_{x\in X}, (\pi_x)_{x\in X}, (b_x)_{x\in X}, (c_x)_{x\in X}]$ is a *μ-continuous tensor product (c.t.p.)-quadruple* if the following conditions are satisfied:

(i) $(\mathcal{H}_x)_{x\in X}$ is a μ-measurable field of Hilbert spaces.
(ii) $(\pi_x)_{x\in X}$ is a μ-measurable field of continuous unitary representations: π_x of G into $\mathcal{H}_x$.
(iii) $(b_x)_{x\in X}$ is a square μ-integrable field of continuous 1-cocycles, such that, for each x in X, b_x is a 1-cocycle of G with respect to π_x.
(iv) $(c_x)_{x\in X}$ is a μ-measurable field of continuous mappings c_x from G into the 1-dimensional torus $\mathbb{T}$, such that, for all γ, γ' in G

$$c_x(\gamma \cdot \gamma') = c_x(\gamma) \cdot c_x(\gamma')\exp\{i \operatorname{Im}\langle b_x(\gamma), \pi_x(\gamma)\, b_x(\gamma')\rangle\}$$

REMARK. Let $[(\mathcal{H}_x)_{x\in X}, (\pi_x)_{x\in X}, (b_x)_{x\in X}, (c_x)_{x\in X}]$ be a μ-c.t.p.-quadruple; from Corollary 1 it follows that for all x in X we get a continuous unitary representation of type (S) of G into $S\mathcal{H}_x$, namely the representation U_{c_x,π_x,b_x}.

Example 8. Let $U_{c,\pi,b}$ be a continuous unitary representation of type (S) of G into some symmetric Hilbert space $S\mathcal{H}$, and for each x in X let us consider $\mathcal{H}_x = \mathcal{H}$, $\pi_x = \pi$, $b_x = b$, and $c_x = c$; the constant fields $(\mathcal{H}_x)_{x\in X}$, $(\pi_x)_{x\in X}$, $(b_x)_{x\in X}$, and $(c_x)_{x\in X}$ satisfy the properties required in Definition 12 in order to get a μ-c.t.p.-quadruple.

Example 9. Let $H = \bigcup_{x\in X} \mathcal{H}_x$ be a C^∞-vector bundle, the fibers $\mathcal{H}_x$ being real Hilbert spaces, let $(\pi_x)_{x\in X}$ be a μ-measurable field of continuous unitary representations π_x of G into $\mathcal{H}_x$, and $(b_x)_{x\in X}$ be a μ-square measurable field of continuous 1-cocycles b_x of G with respect to π_x; let us denote by $\bar{1}_x$, for all x in X, the mapping from G into $\mathbb{T}$ given by $\bar{1}_x(\gamma) = 1\ \forall\, \gamma \in G$. Then $[(\mathcal{H}_x)_{x\in X}, (\Pi_x)_{x\in X}, (b_x)_{x\in X}, (\bar{1}_x)_{x\in X}]$ is a μ-c.t.p.-quadruple, and for each x in X the corresponding unitary representation of type (S) is, by Corollary 3, the representation $\mathrm{EXP}_{b_x}\pi_x$.

Let $[(\mathcal{H}_x)_{x\in X}, (\pi_x)_{x\in X}, (b_x)_{x\in X}, (c_x)_{x\in X}]$ be a μ-c.t.p.-quadruple and let:

$\mathcal{H}_\mu$ be the Hilbert space $\int_X^\oplus \mathcal{H}_x \, d\mu(x)$;
$\tilde{\pi}$ be the continuous unitary representation of $\mathcal{D}(X, G)$ into $\mathcal{H}_\mu$ given for any g in $\mathcal{D}(X, G)$ by

$$\tilde{\pi}(g) = \int_X^\oplus \pi_x(g(x)) \, d\mu(x);$$

$\tilde{b}$ be the continuous mapping from $\mathcal{D}(X, G)$ into $\mathcal{H}_\mu$ such that, for any g in $\mathcal{D}(X, G)$ and for all x in X, $\tilde{b}(g)(x) = b_x(g(x))$;
$\tilde{c}$ be the continuous mappings from $\mathcal{D}(X, G)$ into $\mathbb{T}$ such that, for any g in $\mathcal{D}(X, G)$,

$$\tilde{c}(g) = \int_X c_x(g(x)) \, d\mu(x).$$

One has by an easy computation:
Lemma 5. (i) $\tilde{b}$ is a continuous 1-cocycle of $\mathcal{D}(X, G)$ with respect to $\tilde{\pi}$; moreover, if the b_x, $x \in X$, are trivial 1-cocycles for G, $\tilde{b}$ is a trivial 1-cocycle for $\mathcal{D}(X, G)$.

(ii) For all g, g' in $\mathcal{D}(X, G)$, one has

$$\tilde{c}(g \cdot g') = \tilde{c}(g) \cdot \tilde{c}(g') \cdot \exp\{i \operatorname{Im} \langle \tilde{b}(g), \tilde{\pi}(g)\tilde{b}(g')\rangle\}$$

In particular the triple $(\tilde{\pi}, \tilde{b}, \tilde{c})$ gives rise to a continuous unitary representation of type (S) of $\mathcal{D}(X, G)$, more precisely the representation $U_{\tilde{c},\tilde{\pi},\tilde{b}}$. □
Definition 13. The continuous unitary representation $U_{\tilde{c},\tilde{\pi},\tilde{b}}$ of type (S) of $\mathcal{D}(X, G)$ is called the *continuous tensor product of the μ-c.t.p.-quadruple* constituted by the fields $(\mathcal{H}_x)_{x\in X}, (\pi_x)_{x\in X}, (b_x)_{x\in X}, (c_x)_{x\in X}$.

As an obvious result one gets
Lemma 6. $U_{\tilde{c},\tilde{\pi},\tilde{b}}$ is a nonlocated and continuous unitary representation of order zero of $\mathcal{D}(X, G)$; moreover, each diffeomorphism of X which preserves μ leaves invariant the unitary class of equivalence of $U_{\tilde{c},\tilde{\pi},\tilde{b}}$.

2.5.2 Basic Properties of $U_{\tilde{c},\tilde{\pi},\tilde{b}}$

Proposition 8. If, for all x in X, $b_x(G)$ is a total set in $\mathcal{H}_x$, then the vacuum vector EXP 0 in $S\mathcal{H}_\mu$ is cyclic for $U_{\tilde{c},\tilde{\pi},\tilde{b}}$.

PROOF. Let F be the closed subspace of $S\mathcal{H}_\mu$ spanned by the set $\{U_{\tilde{c},\tilde{\pi},\tilde{b}}(g) \text{ EXP } 0, g \in \mathcal{D}(X, G)\}$ and let Θ be the complete Boolean algebra spanned by the μ-measurable subsets of X. For an element θ of Θ we shall denote by $\mathcal{H}^\theta$ the Hilbert space $\int_\theta^\oplus \mathcal{H}_x \, d\mu(x)$, by P_θ the orthogonal projection from $\mathcal{H}_\mu$ onto $\mathcal{H}^\theta$, and by F_θ the closed subspace of $S\mathcal{H}^\theta$ spanned by $\{\text{EXP}(P_\theta(\tilde{b}(g))), g \in \mathcal{D}(X, G)\}$; we denote by $\omega_\theta = \text{EXP } 0_\theta$ the vacuum vector of $S\mathcal{H}^\theta$.

For any element g of $\mathcal{D}(X, G)$, $P_\theta(\tilde{b}(g)) = \tilde{b}(g')$ for some g' in $\mathcal{D}(X, G)$ with support in θ; it follows that a family $\Phi = (\theta_i)_{i \in I}$ of elements of Θ being given, where I is countable, and $\theta_i \cap \theta_j = \emptyset$ if $i \neq j$, one gets a canonical isomorphism Λ_Φ from the countable tensor product $\bigotimes_{i \in I} F_{\theta_i}$ (based on the ω_{θ_i}) into F_θ which sends $\bigotimes_{i \in I} \omega_{\theta_i}$ into ω_θ, where $\theta = \bigcup_{i \in I} \theta_i$.

One easily sees that the pairs $(F_\theta, \omega_\theta)$, θ in Θ, and the isomorphisms Λ_Φ as above, give rise to a complete Boolean algebra of tensor decompositions of $(F, \text{EXP } 0)$. Moreover, $b_x(G)$ being total in the $\mathcal{H}_x$, for all x in X, it follows that the set $\{\text{EXP } \tilde{b}(g), g \in \mathcal{D}(X, G)\}$ is total in F, and then the set of factorizable vectors ξ satisfying $\langle \text{EXP } 0, \xi \rangle = 1$ is total in F. Owing to the Araki-Woods theorem [3, Theorem 6.1], there exists a Hilbertian integral $H = \int_X^\oplus H_x \, d\mu(x)$ and an isomorphism of Hilbert spaces $\lambda: F \to SH$ with $\lambda(\text{EXP } 0) = \text{EXP } 0$ which sends the factorizable vectors of F into the ones of SH. It follows that we can define, for all g in $\mathcal{D}(X, G)$, the element $B(g)$ of H by $\text{EXP}(B(g)) = \lambda(\text{EXP}(\tilde{b}(g)))$; since the set $\{\text{EXP}(\tilde{b}(g)) | g \in \mathcal{D}(X, G)\}$ is total in F, it follows that $\{\text{EXP } B(g) | g \in \mathcal{D}(X, G)\}$ is total in SH, and then $\{B(g) | g \in \mathcal{D}(X, G)\}$ is total in H.

From the fact that $\exp a = \exp a'$ implies $a = a' (\text{mod } 2i\pi)$, it follows that, for all g, g' in $\mathcal{D}(X, G)$, one has $\langle B(g), B(g') \rangle = \langle \tilde{b}(g), \tilde{b}(g') \rangle \text{ mod } 2i\pi$. For θ in Θ, let P_θ be the orthogonal projection from $\mathcal{H}_\mu$ onto $\mathcal{H}^\theta$, and let P'_θ be the orthogonal projection from H onto the subspace $\int_\theta^\oplus \mathcal{H}_x \, d\mu(x)$; one has, owing to the factorizability of λ,

$$\varphi(\theta) = (2i\pi)^{-1}\{\langle P'_\theta B(g), B(g') \rangle - \langle P_\theta \tilde{b}(g), \tilde{b}(g') \rangle\}$$

which is an integer; so we get a σ-additive mapping $\varphi: \theta \to \varphi(\theta)$ from Θ into the ring $\mathbb{Z}$ integers. But from [10, Appendix C, Lemma C.2], we can find a partition of the unity $(\theta_i)_{i \in I}$ of Θ such that, for all i in I, $|\varphi(\theta_i)| < 1/2$; as φ takes integer values, it follows that φ is zero everywhere, and then

$$\langle B(g), B(g') \rangle = \langle \tilde{b}(g), \tilde{b}(g') \rangle \qquad \text{for all } g, g' \text{ in } \mathcal{D}(X, G)$$

As $B(\mathcal{D}(X, G))$ and $\tilde{b}(\mathcal{D}(X, G))$ are total sets in H and respectively $\mathcal{H}_\mu$, we can find a unitary operator $U: H \to \mathcal{H}_\mu$ such that $U(B(g)) = \tilde{b}(g)$ for all g in $\mathcal{D}(X, G)$, which gives the unitary operator $SU: SH \to S\mathcal{H}_\mu$ such that for all h in H, $SU(\text{EXP } h) = \text{EXP}(Uh)$; let us consider now $SU \circ \lambda$: it is a unitary operator from F onto $S\mathcal{H}_\mu$ which leaves invariant the elements of the form $\text{EXP } h$, so $SU \circ \lambda$ is the canonical injection.

In order to study the irreducibility or the reducibility of $U_{\tilde{c},\tilde{\pi},\tilde{b}}$, we have to compute its commutant. This is a highly nontrivial work which was done by Delorme [7]: using the theory of standard forms of von Neumann algebras and a theorem of Araki and Woods on the complete Boolean algebras of factors of type 1, Delorme proved [7, Theorem 1] the following

Proposition 9. The commutant of $U_{\tilde{c},\tilde{\pi},\tilde{b}}$ is spanned by the unitary operators of the form $U_{\gamma,A,\beta}$, where A is decomposable in $\int_X^\oplus \mathcal{H}_x \, d\mu(x)$.

2.5.3 Multiplicative Integrals in the Case $G = SO(n, 1)$, $SU(n, 1)$, or a Motion Group

An important theorem about the irreducibility of some continuous tensor products is the following [7, Theorem 3].

Proposition 10. Let X be endowed with a strictly positive and nonatomic measure μ, let G be a Lie group, and let $[(\mathcal{H}_x)_{x\in X}, (\pi_x)_{x\in X}, (b_x)_{x\in X}, (c_x)_{x\in X}]$ be a μ-c.t.p.-quadruple with the following property: there exists a compact subgroup K of G such that, μ almost everywhere:

(a) the restriction of π_x to K does not contain the trivial representation of K.

(b) b_x is zero on K, and $b_x(G)$ is total in $\mathcal{H}_x$.

Then $U_{\tilde{c},\tilde{\pi},\tilde{b}}$ is irreducible.

PROOF. Let $(T_x)_{x\in X}$ be a μ-measurable field of unitary operators T_x on $\mathcal{H}_x$, and let $(\beta_x)_{x\in X}$ be a μ-square integrable field of vectors β_x in $\mathcal{H}_x$; from Proposition 9 it suffices to prove that any unitary operator of the form $U_{1,T,\beta}$, with $T = \int_X^\oplus T_x \, d\mu(x)$ and $\beta = \int_X^\oplus \beta_x \, d\mu$, which commutes with $U_{\tilde{c},\tilde{\pi},\tilde{b}}$ is a scalar operator, and more precisely the identity $\mathbb{1}_{S\mathcal{H}_\mu}$ on $S\mathcal{H}_\mu$. An easy computation shows that $U_{1,T,\beta} \cdot U_{\tilde{c},\tilde{\pi},\tilde{b}} = U_{\tilde{c},\tilde{\pi},\tilde{b}} \cdot U_{1,T,\beta}$ implies, for μ-almost every x in X, that

$$T_x \cdot \pi_x(g) = \pi_x(g) \cdot T_x \qquad \text{and}$$

$$\beta_x + T_x \cdot b_x(g) = b_x(g) + \pi_x(g)\beta_x \tag{*}$$

for any g in G. In particular, for any g in K, one gets $\beta_x = \pi_x(g)\beta_x$, from which it follows that $\beta = 0$. Then (*) becomes $T_x \cdot b_x(g) = b_x(g)$ for all g in G, and as $b_x(G)$ is total in $\mathcal{H}_x$ it follows that T_x is the identity on $\mathcal{H}_x$ for μ-almost every x in X, and then $U_{1,T,\beta}$ is necessarily equal to $\mathbb{1}_{S\mathcal{H}_\mu}$.

As a corollary one gets the following result which was given without proof in [26, Theorem 1], and which can be easily applied to the case where G is one of the groups $SO(n, 1)$, $SU(n, 1)$, or a motion group.

Corollary. Let π be a continuous unitary representation of G into a real Hilbert space $\mathcal{H}^0$, let b be a continuous 1-cocycle of G with respect to π, and let μ be a nonatomic positive measure on X. For all x in X, let $\mathcal{H}_x^0 = \mathcal{H}^0$, $\pi_x = \pi$, and $b_x = b$. If there is a compact subgroup K of G such that $b \mid K = 0$, $b(G)$ being total in $\mathcal{H}^0$, and such that $\mathcal{H}^0$ contains no nonzero vector invariant with respect to K, then the c.t.p.u.r. $\mathrm{EXP}_{\tilde{b}}\tilde{\pi} = U_{\tilde{1},\tilde{\pi},\tilde{b}}$ is irreducible.

REMARKS. (1) The c.t.p.u.r. $U_{\tilde{c},\tilde{\pi},\tilde{b}}$ satisfying Proposition 10 or its corollary, when K is not reduced to $\{e\}$, are necessarily such that for μ-almost every x in X, π_x cannot be the trivial representation of G, and b_x is a nontrivial 1-cocycle for π_x. The case where π_x is trivial for all x will be studied in the next section.

(2) Among the semisimple Lie groups with simple Lie algebra, the only groups with nontrivial unitary cohomology with respect to an irreducible unitary representation are $SO(n; 1)$ and $SU(n; 1)$, by Lemma V.5 and Theorem V.5 of [6]. In particular, if G is a compact semisimple Lie group, it is not possible to find a c.t.p.u.r. satisfying the conditions of Proposition 10. The explicit form of the pair (π, b) satisfying the conditions of the corollary above, and therefore giving a multiplicative integral with respect to the measure μ on X, is given in [26] in both cases $G = SO(n; 1)$ and $G = SU(n; 1)$.

1. In the case $G = SO(n; 1)$, let $\mathbb{R}^n$ be endowed with the scalar product denoted $\langle\ ,\ \rangle$, let S_{n-1} be the sphere $\{\omega = (\omega_1, \omega_2, \ldots, \omega_n) \in \mathbb{R}^n \mid \langle \omega, \omega\rangle = 1\}$ endowed with an invariant measure λ, and let $\mathcal{H}^0$ be the Hilbert space spanned by the real continuous function φ on S_{n-1} satisfying $\int_{S_{n-1}} \varphi(\omega)\, d\lambda(\omega) = 0$, with respect to the Euclidean norm on the sphere:

$$\|\varphi\| = \left[\int_{S_{n-1}\times S_{n-1}} \ln\,(1 - \langle\omega, \omega'\rangle)\varphi(\omega)\varphi(\omega')\, d\lambda(\omega)\, d\lambda(\omega')\right]^{1/2}$$

An element $g = (g_{ij})_{1\le i,j\le n+1}$ in G acts on S_{n-1} as follows: for $\omega = (\omega_1, \omega_2, \ldots, \omega_n)$ in S_{n-1}, $g \cdot \omega = (\omega_1', \omega_2', \ldots, \omega_n')$

with

$$\omega_i' = (\omega_1 g_{1,i} + \cdots + \omega_n g_{n,i} + g_{n+1,i}) \times (\omega_1 g_{1,n+1} + \cdots + \omega_n g_{n,n+1} + g_{n+1,n+1})^{-1}$$

One gets a continuous unitary representation π of $SO(n; 1)$ into $\mathcal{H}^0$ given by

$$(\pi(g)\varphi)(\omega) = |\omega_1 g_{1,n+1} + \cdots + \omega_n g_{n,n+1} + g_{n+1,n+1}|^{1-n}\varphi(g \cdot \omega)$$

$g \in SO(n; 1)$, $\varphi \in \mathcal{H}^0$, $\omega \in S_{n-1}$. The corresponding nontrivial 1-cocycle b is given, for all g in $SO(n; 1)$ by

$$b(g)(\omega) = |\omega_1 g_{1,n+1} + \cdots + \omega_n g_{n,n+1} + g_{n+1,n+1}|^{1-n} - 1$$

A rather long, but straightforward computation shows that the pair (π, b) satisfies the conditions of the corollary of Proposition 10, and then the corresponding c.t.p.u.r. $\text{EXP}_{\tilde{b}}\tilde{\pi}$ is in the class of a multiplicative integral. The case $G = SU(n; 1)$ is quite similar (see [26, Example b]).

2. A very simple example is provided by the motion groups. Let G be a motion group, i.e., G is the semidirect product $K \cdot V$ of a compact Lie group K with a finite-dimensional real vector space V, K acting on V by $(k, v) \to \tau(k)v$, where τ is a representation of K into V. The product in $K \cdot V$ is then given by

$$(k, v) \cdot (k', v') = (kk', v + \tau(k)v')$$

for all k, k' in K and all v, v' in V.

Let θ be an irreducible subrepresentation of τ in some nontrivial subspace W of V. As K is compact, we can endow V with a scalar product for which τ and θ are unitary representations; we shall denote by p the orthogonal projection from V onto W. One gets an irreducible continuous representation π of $K \cdot V$ into W by $\pi(k, v) = \theta(k)$, $(k, v) \in K \cdot V$. Now, let us consider the mapping b: $K \cdot V \to W$ given by $b(k, v) = p(v)$, $(k, v) \in K \cdot V$. For any two elements (k, v) and (k', v') of $K \cdot V$ one has

$$\begin{aligned} b[(k, v) \cdot (k', v')] &= b(kk', v + \tau(k)v') \\ &= p(v + \tau(k)v') = p(v) + p \cdot \tau(k)v' \end{aligned}$$

As the orthogonal complement of W in V is also invariant with respect to τ, it follows that p commutes with the operators $\tau(k)$. One easily deduces that $b[(k, v) \cdot (k', v')] = b(k, v) + \pi(k, v)b(k', v')$, and then b is a continuous 1-cocycle of $G = K \cdot V$ with respect to π. One has $b(K) = 0$, $b(G) = W$; moreover, owing to the irreducibility of θ, a nonzero vector of W cannot be invariant under the operator $\theta(k)$. It follows that the pair (π, b) satisfies the conditions of the corollary of Proposition 10, and then gives rise to an irreducible c.t.p.u.r., the representation $\text{EXP}_{\tilde{b}}\tilde{\pi}$.

2.6 THE CASE WHERE G IS A NILPOTENT LIE GROUP

2.6.1 Multiplicative Integrals for Nilpotent Lie Groups

In this section we suppose that G is a real connected and simply connected nilpotent Lie group with Lie algebra $\mathcal{G}$; L: $G \to \mathcal{G}$ will denote the inverse of the exponential mapping from $\mathcal{G}$ onto G, $\mathcal{G}'$ will denote the dual of $\mathcal{G}$. We shall set $\mathcal{G}^{(1)} = [\mathcal{G}, \mathcal{G}]$ and $\mathcal{G}^{(2)} = [\mathcal{G}, \mathcal{G}^{(1)}]$, and we shall consider the set

$$\tilde{\mathcal{G}}^1 = \{\lambda \in \mathcal{G}' | \lambda(\mathcal{G}^{(1)}) \neq 0, \lambda(\mathcal{G}^{(2)}) = (0)\}$$

Each element α of $\tilde{\mathcal{G}}'$ gives rise to the ideal

$$\mathcal{G}(\alpha) = \{u \in \mathcal{G} | \alpha([u, \mathcal{G}]) = 0\}$$

which is the kernel of the bilinear and skew-symmetric form B_α: $(u, v) \to \alpha([u, v])$. It follows that the space $V_\alpha = \mathcal{G}/\mathcal{G}(\alpha)$ is a real vector space of even dimension, on which B_α induces a symplectic form $\tilde{B}_\alpha$; the choice of a symplectic basis with respect to $\tilde{B}_\alpha$ allows us to endow V_α with a complex Hilbertian structure according to some scalar product denoted by $\langle\ ,\ \rangle_\alpha$; we shall denote by I_α the trivial representation of G into V_α, and by p_α the canonical projection from $\mathcal{G}$ onto V_α.

Lemma 7. Let α be in $\tilde{\mathcal{G}}'$; the mapping $b^\alpha = p_\alpha \circ L$ is a smooth homomorphism from G onto V_α, hence a continuous nontrivial 1-cocycle of G with respect to I_α. Moreover the mapping $c^\alpha: g \to \exp\{i\alpha(L(g))\}$ is a smooth mapping from G into $\mathbb{T}$, such that, for all g, g' in G,

$$c^\alpha(g \cdot g') = c^\alpha(g) \cdot c^\alpha(g') \cdot \exp\{i \operatorname{Im}\langle b^\alpha(g), b^\alpha(g')\rangle_\alpha\}$$

PROOF. As α is a linear form on a finite-dimensional space, as p_α is a smooth mapping onto V_α, and as L is a bijective smooth mapping, it follows that b^α is smooth and onto V_α and that c^α is smooth too. Let $g = \exp u$, $g' = \exp u'$ be two elements of G, with u, u' in $\mathcal{G}$. The Campbell-Hausdorff formula (see, e.g., [4]) allows us to write

$$\begin{aligned} g \cdot g' &= \exp u \cdot \exp u' \\ &= \exp(u + u' + \theta(u, u')) \qquad \text{with } \theta(u, u') \text{ in } \mathcal{G}^{(1)} \end{aligned}$$

which is contained in $\mathcal{G}(\alpha)$, and then

$$(p_\alpha \circ L)(\exp u \cdot \exp u') = p_\alpha(u + u' + \theta(u, u')) = p_\alpha(u) + p_\alpha(u')$$

that is, $b^\alpha(g \cdot g') = b^\alpha(g) + b^\alpha(g')$. Moreover,

$$c^\alpha(g \cdot g') = \exp\{i\alpha(u + u' + \theta(u, u'))\}$$

As (see [4])

$$\theta(u, u') = \tfrac{1}{2}[u, u'] + \theta_1(u, u') \qquad \text{with } \theta_1(u, u') \text{ in } \mathcal{G}^{(2)},$$

one has

$$c^\alpha(gg') = \exp\{i\alpha(u + u' + \tfrac{1}{2}[u, u'])\} = e^{i\alpha(u)} \cdot e^{i\alpha(u')} \cdot e^{i/2\tilde{B}_\alpha(u,u')}$$
$$= c^\alpha(g) \cdot c^\alpha(g') \cdot \exp\{^{i/2}\tilde{B}_\alpha(p_\alpha(u), p_\alpha(v))\}$$

the result follows from $\tilde{B}_\alpha(p_\alpha(u), p_\alpha(u')) = 2\,\mathrm{Im}\langle b^\alpha(g), b^\alpha(g')\rangle_\alpha$.

From Lemmas 3 and 7 it follows that the triple $(c^\alpha, I_\alpha, b^\alpha)$ gives rise to a unitary representation $U^\alpha = U_{c^\alpha, I_\alpha, b^\alpha}$ of type (S) of G into SV_α.

Now, for any x in the manifold X endowed with a nonatomic positive measure μ, let $V_\alpha^x = V_\alpha$, $I_\alpha^x = I_\alpha$, $b_x^\alpha = b^\alpha$, and $c_x^\alpha = c^\alpha$; we get the μ-c.t.p.-quadruple $[(V_\alpha^x)_{x\in X}, (I_\alpha^x)_{x\in X}, (b_x^\alpha)_{x\in X}, (c_x^\alpha)_{x\in X}]$ which yields the c.t.p.u.r. $U_\mu^\alpha = U_{\tilde{c}^\alpha, \tilde{I}_\alpha, \tilde{b}^\alpha}$ on SV_α^μ, with

$$V_\alpha^\mu = \int_X^\oplus V_\alpha^x \, d\mu(x), \qquad \tilde{I}_\alpha = \int_X^\oplus I_\alpha^x \, d\mu(x)$$

$$\tilde{b}^\alpha = \int_X^\oplus b_x^\alpha \, d\mu(x), \qquad \tilde{c}^\alpha : g \to \int_X c_x^\alpha(g(x)) \, d\mu(x)$$

(cf. Lemma 6), which is invariant by all the diffeomorphisms of X preserving μ.

We have the following result.

Proposition 11. Let X be endowed with a nonatomic positive measure μ, and let G be a connected and simply connected nilpotent Lie group with Lie algebra $\mathcal{G}$. For each α in $\tilde{\mathcal{G}}$, the c.t.p.u.r. U_μ^α is in the class of a multiplicative G-integral with respect to μ.

PROOF. It suffices to prove the irreducibility of U_μ^α. As G is a connected nilpotent Lie group, one knows that G is of type 1 and that $H^1(G, \pi) = (0)$ for all nontrivial irreducible unitary representations π (see, e.g., [12]). But for such a group Delorme has shown that U_μ^α is irreducible if and only if $U_{c^\alpha, I_\alpha, b^\alpha}$ is irreducible [5, Theorem III-2]; the assertion follows from the fact that the representation $U_{c^\alpha, I_\alpha, b^\alpha}$, $\alpha \in \tilde{\mathcal{G}}'$ is known to be irreducible [5, Lemma IV.2].

2.6.2 Unitary Characters of $\mathcal{D}$(X, G); Representations of Order >0

Let α be in $\tilde{\mathcal{G}}'$; from Lemma 7, it follows that, for all g in $\mathcal{D}(X, G)$, $\tilde{b}^\alpha(g)$: $x \to b^\alpha(g(x))$ is an element of $\mathcal{D}(X, V_\alpha)$, and that $\tilde{b}^\alpha : g \to \tilde{b}^\alpha(g)$ is a continuous homomorphism from $\mathcal{D}(X, G)$ into $\mathcal{D}(X, V_\alpha)$.

Let us consider the multidistributions space $\mathcal{D}'_{\mathbb{R}}(X, V_\alpha)$ consisting of all the real-valued elements of the dual space of $\mathcal{D}(X, V_\alpha)$. From the above discussion one sees easily that

Lemma 8. For all α in $\tilde{\mathcal{G}}'$ and all Φ in $\mathcal{D}'_{\mathbb{R}}(X, V_\alpha)$, the mapping $\chi^\alpha_\Phi: g \to \exp\{i\langle\Phi, \tilde{b}^\alpha(g)\rangle\}$ is a continuous unitary character of $\mathcal{D}(X, G)$, the support and the order of χ^α_Φ being the support and the order of Φ.

We are now prepared to study infinite-dimensional irreducible unitary representations of $\mathcal{D}(X, G)$ of any order and with support X.

Proposition 12 Let G be a connected and simply connected nilpotent Lie group with Lie algebra $\mathcal{G}$, and let X be endowed with a positive nonatomic measure μ with support X. A triple $(\alpha_1, \alpha_2, \Phi)$ being given, with (α_1, α_2) in $\tilde{\mathcal{G}}' \times \tilde{\mathcal{G}}'$ and Φ in $\mathcal{D}'_{\mathbb{R}}(X, V_{\alpha_2})$, the mapping $\Theta^{\mu,\alpha_1}_{\alpha_2,\Phi}$ given for all g in $\mathcal{D}(X, G)$ by

$$\Theta^{\mu,\alpha_1}_{\alpha_2,\Phi}(g) = \chi^{\alpha_2}_\Phi(g) \otimes U^{\alpha_1}_\mu(g)$$

is a continuous irreducible unitary representation of type (S) of $\mathcal{D}(X, G)$ with support X, and with order the order of Φ.

PROOF. From Proposition 10 and Lemma 8, the only thing we have to do is to prove that $\Theta^{\mu,\alpha_1}_{\alpha_2,\Phi}$ is a unitary representation of type (S). For all g in $\mathcal{D}(X, G)$ one has

$$\Theta^{\mu,\alpha_1}_{\alpha_2,\Phi}(g) = e^{i\langle\Phi,\tilde{b}^{\alpha_2}(g)\rangle}\, U_{\tilde{c}^{\alpha_1},\tilde{I}_{\alpha_1},\tilde{b}^{\alpha_1}}(g) = U_{\chi^{\alpha_2}_\Phi\cdot\tilde{c}^{\alpha_1},\tilde{I}_{\alpha_1},\tilde{b}^{\alpha_1}}(g)$$

In order to prove that $\Theta^{\mu,\alpha_1}_{\alpha_2,\Phi}$ is of type (S), from Corollary 1 it suffices to prove that, for all g, g' in $\mathcal{D}(X, G)$,

$$(\chi^{\alpha_2}_\Phi \cdot \tilde{c}^{\alpha_1})(gg') \cdot (\chi^{\alpha_2}_\Phi \cdot \tilde{c}^{\alpha_1}(g))^{-1} \cdot (\chi^{\alpha_2}_\Phi \cdot \tilde{c}^{\alpha_1}(g'))^{-1}$$
$$= \exp\{i \operatorname{Im}\langle\tilde{b}^{\alpha_1}(g), \tilde{b}^{\alpha_1}(g')\rangle\}$$

but this condition is satisfied owing to the fact that $\chi^{\alpha_2}_\Phi$ is a unitary character and to the fact that, by Lemma 7,

$$\tilde{c}^{\alpha_1}(gg') \cdot \tilde{c}^{\alpha_1}(g)^{-1} \cdot \tilde{c}^{\alpha_1}(g')^{-1} = \exp\{i \operatorname{Im}\langle\tilde{b}^{\alpha_1}(g), \tilde{b}^{\alpha_1}(g')\rangle\} \qquad \square$$

COMMENTS. (a) In the case $G = \mathbb{R}^n$, $\hat{\mathcal{D}}(X, G)$ is completely known: by Section 2.2.2 one easily deduces that the elements of $\hat{\mathcal{D}}(X, \mathbb{R}^n)$ are of the form $\chi_T: g \to \exp(i\langle T, g\rangle)$ where T is an element of $\mathcal{D}'_{\mathbb{R}}(X, \mathbb{R}^n)$. If G is nilpotent, Proposition 12 gives many G-distributions of any order, and it seems reasonable to conjecture that Proposition 12 gives all the nonlocated G-distributions.

(b) The case G semisimple is far from being solved. In particular, if G is semisimple noncompact, $G \neq SL(2, IR)$, $G \neq SO(n; 1)$, $G \neq SU(n; 1)$, nothing seems to be known of the existence of nonlocated G-distributions. The case where G is a compact semisimple Lie group will be studied in the next chapter.

REFERENCES

1. S. Albeverio and B. Torresani. *Some remarks on representations of jet groups and gauge groups*, Marseille preprint (1992).
2. H. Araki, *Publ. R.I.M.S., Kyoto Univ.*, *5*:361 (1970).
3. H. Araki and E. J. Woods, *Publ. R.I.M.S., Kyoto Univ.*, *2*:157 (1966).
4. N. Bourbaki, *Groupes et Algèbres de Lie*, Hermann, Paris, Chap. 2–3 (1972).
5. P. Delorme, *1-cohomologie et produits tensoriels continus de représentations*, Thèse, 3ème cycle, Paris (1975).
6. P. Delorme, *Bull. Soc. Math. France*, *105*:291 (1977).
7. P. Delorme, *J. Funct. Anal.*, *30*:36 (1978).
8. J. Dixmier, *Les C* Algèbres et leurs Représentations*, Gauthier-Villars, Paris (1969).
9. I. M. Gelfand and M. I. Graev, *Funkt. Anal. i Prilozhen*, *2*(1):20 (1968). English transl. *Funct. Anal. Appl.*, *2*:19 (1968).
10. I. M. Gelfand and N. Y. Vilenkin, *Les Distributions*. Vol. 4, Dunod, Paris (1967); *Generalized Functions*, Vol. 4, *Applications of Harmonic Analysis*, Academic Press, New York (1964). (Translation of Russian edition of 1961).
11. A. Guichardet, *Symmetric Hilbert Spaces and Related Topics*, Lecture Notes in Math., vol. 26, Springer-Verlag, Berlin, Heidelberg, New York (1972).
12. A. Guichardet, *Bull. Soc. Math. France*, *96*:305 (1972).
13. A. Guichardet, "*Représentations de G^X selon Gelfand et Delorme*," Séminaire Bourbaki, n°.486, Paris (1976).
14. A. Guichardet, *Math. Ann.*, *228*:215 (1977).
15. A. Guichardet and A. Wulfsohn, *J. Funct. Anal.*, *2*:371 (1968).
16. T. Hida, *Brownian Motion*, Springer, Berlin (1980).
17. D. A. Kazhdan, *Funkt. Anal. i Prilozhen*, *1*:71 (1967); English transl., *Funct. Anal. Appl.*, *1*:63 (1967).
18. S. I. Karpushev, *Vestnik Leningrad, Univ. Math. Mekh.*, *1981*, 37; English transl., *SIAM J. Control Optim.*, *19*:37 (1981).
19. G. W. Mackey, *Acta Math.*, *9*:265 (1958).
20. J. Marion, *Anal. Pol. Math.*, *43*:71 (1984).
21. J. Marion, *1-Cohomology of gauge groups and partial gauge actions according to the derivatives of the Maurer-Cartan cocycle*, preprint, Z.I.F. Bielefeld University (1983).
22. K. R. Parthasarathy, *Commun. Math. Phys.*, *16*:148 (1970).
23. L. Schwartz, *Théorie des Distributions*, Hermann, Paris (1966).
24. R. F. Streater, *Rend. Sci. Int. Fis. E. Fermi*, *no. 11*:247 (1969).
25. A. M. Vershik, I. M. Gelfand, and M. I. Graev, *Usp. Math. Naut.*, *28*(5):83 (1973); English transl., *Russ. Math. Surv.* *28*(5):87 (1973).
26. A. M. Vershik, I. M. Gelfand, and M. I. Graev, *Funkt. Anal. i Pril*, *8*:67 (1974); English transl., *Funct. Anal. Appl.*, *8*:151 (1974).
27. A. M. Vershik and S. I. Karpushev, *Math. USSR-Sbornik* *47*:513 (1984) (transl.).
28. S. P. Wang, *Proc. Amer. Math. Soc.*, *42*:621 (1974).
29. D. X. Xia, *Measure and Integration Theory on Infinite-Dimensional Spaces*, Academic Press, New York (1972).

3

The Energy Representations of Gauge Groups

3.1 INTRODUCTION

This chapter is devoted to the construction and the study of a class of nonlocated and order 1 unitary representations of $\mathcal{D}(X, G)$ when G is a compact semisimple Lie group. In this case $\mathcal{D}(X, G)$ is called a gauge group; as already mentioned in the introduction, it plays a prominent part in mathematical physics and theoretical physics, in particular in gauge fields theories and noncommutative generalizations of random field theories (see, e.g., [2-5, 14]). Unfortunately, as mentioned in Chapter 2, Section 4, compact Lie groups have a trivial 1-cohomology with respect to any unitary representation. It follows then, from the corollary of Lemma 4 in Section 3 of Chapter 2, that any continuous tensor product of unitary representations of G gives necessarily a reducible representation of $\mathcal{D}(X, G)$.

The representation V of $\mathcal{D}(X, G)$ on the space $\mathcal{D}_1(X, \mathcal{G})$ of $\mathcal{G}$-valued and compactly supported smooth 1-forms on X with $\mathcal{G}$ the Lie algebra to G has a nontrivial 1-cocycle, the so-called Maurer-Cartan cocycle $b: g \to b(g) = dg \cdot g^{-1}$ (Section 9, Chapter 1). As a natural consequence, we can make use of a general process in order to get unitary representations of type (S) of $\mathcal{D}(X, G)$, which are nonlocated, of order 1, and which may be irreducible. More precisely let us assume that there is an orthogonal representation π of $\mathcal{D}(X, G)$ on some real pre-Hilbertian space E and a continuous linear mapping p from $\mathcal{D}_1(X, \mathcal{G})$ onto E such that, for all g in $\mathcal{D}(X, G)$, $p \cdot V(g) = \pi(g) \cdot p$. One easily sees that $p \cdot b$ is a nontrivial 1-cocycle of $\mathcal{D}(X, G)$ with respect to the unitary representation which extends π on the Hilbert completion of E. Consequently each triple (π, E, p) satisfying the above assumption gives rise, by the process described in Chapter 2, Section 3 (i.e., by specifying a positive definite function) to a unitary representation of type (S) of $\mathcal{D}(X, G)$, which can be realized on the Fock space $SH_{\mathbb{C}}$, where $H_{\mathbb{C}}$ denotes the complexified Hilbert completion of E.

Notice that for such a representation of type (S) based on $(\pi, p \cdot b)$ the corresponding spherical function φ associated with the vacuum vector in $SH_{\mathbb{C}}$ is given by

$$\varphi(g) = \exp[-\tfrac{1}{2}\|p(dg \cdot g^{-1})\|^2], \qquad g \in \mathcal{D}(X, G) \tag{*}$$

The pioneer's works which used this process [3, 10, 22, 23], as well as the first papers which came after [7, 15] were concerned with the case $E = \mathcal{D}_1(X, \mathcal{G})$, $\pi = V$, p being the identity operator, and the invariant Euclidean norm on $\mathcal{D}_1(X, \mathcal{G})$ being the L^2-norm associated with the choice of a smooth Riemannian structure on X and to the opposite of the Killing form on $\mathcal{G}$. In this case, φ looks like a noncommutative and multiplicative version of an integral of energy; this explains the fact that these representations were called [3], as well as their later generalizations [15–18], "energy representations."

The main part of this chapter is devoted to the construction and the study of what we call "basic energy representations," which are exactly the energy representations of $\mathcal{D}(X, G)$ associated to the Riemannian flags of X. These representations are obtained by the general process described above; they constitute the natural generalization of the ones given in [3, 10, 22, 23]. The study of their reducibility or irreducibility is rather difficult, using deep results [1, 2, 4] about Gaussian measures and dilations and regularizations of a certain random field connected with the Dirichlet form associated to the selected Riemannian flag.

The last part of the chapter is devoted to the construction of rings of nonlocated and order 1 unitary representations of $\mathcal{D}(X, G)$ which were obtained for the first time in [18]. More precisely, with each Riemannian flag of X is associated a ring $R(F)$ (i.e., a set of representations closed under direct sum and tensor product) which is, roughly speaking, parametrized by the union of all the unitary duals of the symmetric groups S_n, n running over the set of positive integers; moreover the basic energy representation associated with F appears as the element of $R(F)$ corresponding to the unit representation of S_1.

3.2 THE BASIC ENERGY REPRESENTATION ASSOCIATED WITH A RIEMANNIAN FLAG

3.2.1 Riemannian Flag

Definition 1. Let X be a smooth manifold with finite dimension d_0. A *Riemannian flag* F of X is a finite collection $F = (X_k, r_k)_{0 \leq k \leq p}$, where p is an integer satisfying $0 \leq p \leq d_0$, such that

(i) For each integer k with $0 \leq k \leq p$, X_k is a submanifold of X endowed with a Riemannian structure r_k of class C^1.
(ii) For each pair (k, k') of integers such that $0 \leq k \leq k' \leq p$ one has $X_p \subset \cdots \subset X_{k'} \subset \cdots \subset X_k \subset \cdots \subset X_0 = X$.
(iii) For all k such that $0 \leq k \leq p$, the dimension d_k of X_p is equal to $d_0 - k$.

Notice that for a given flag $F = (X_k, r_k)_{0 \leq k \leq p}$ and for a given integer q such that $0 \leq q \leq p$, the family $F^{\leq q} = (X_k, r_k)_{0 \leq k \leq q}$ is also a Riemannian flag; we shall say that $F^{\leq q}$ is a *subflag* of F.

REMARKS. (1) In a Riemannian flag $F = (X_k, r_k)_{0 \leq k \leq p}$ with $p > 0$, for two integers k and k' such that $0 \leq k < k' \leq p$, the Riemannian structure $r_{k'}$ selected on $X_{k'}$ has generally no connection with the induced Riemannian structure on $X_{k'}$ coming from r_k.

(2) (X, r), where r is C^1-Riemannian structure on X, is a flag of X.

(3) Let $F = (X_k, r_k)_{0 \leq k \leq p}$ be a Riemannian flag of X; for any nonempty open subset U of X, the family $F \cap U = (X_k \cap U, r_k)_{0 \leq k \leq p}$ is a Riemannian flag of U.

3.2.2 Energy Scalar Products Associated with a Riemannian Flag

Let $F = (X_k, r_k)_{0 \leq k \leq p}$ be a Riemannian flag of X. For each integer k, $0 \leq k \leq p$, the Riemannian structure r_k gives rise canonically to a volume measure v_k on X_k and to a Euclidean structure on each fiber $T_x X_k$, $x \in X_k$, of the tangent bundle TX_k of X_k.

Let us consider now a finite-dimensional Euclidean (real) vector space E. The nuclear space $\mathfrak{D}_1(X, E)$ of all the E-valued and compactly supported smooth 1-forms on X can be endowed with a real pre-hilbertian structure given by the positive definite inner product $(\ ,\ )_F$ such that, for all ω, ω' in $\mathfrak{D}_1(X, E)$

$$(\omega, \omega')_F = \sum_{k=0}^{p} \int_{X_k} \mathrm{tr}[\omega_k'^*(x) \cdot \omega_k(x)]\, dv_k(x)$$

where, for all x in X_k, $\omega_k(x)$ and $\omega_k'(x)$ denote the restriction of $\omega(x)$ and $\omega'(x)$ to the tangent space T_xX_k of X_k at x, $\omega_k'^*(x)$ denoting the adjoint of $\omega_k'(x)$ with respect to the Euclidean structures of T_xX_k (given by r_k) and of E; tr denotes the trace operator.

Definition 2. The scalar product $(\ ,\ )_F$ defined as above will be called the *energy scalar product* associated with the Riemannian flag F.

We shall denote by H_F the Hilbert completion of $\mathfrak{D}_1(X, E) \otimes \mathbb{C}$ with respect to the sesquilinear extension of $(\ ,\ )_F$ (that we shall denote in the same manner).

3.2.3 The Basic Energy Representation U_F

Let F be a Riemannian flag of X, and let G be a compact semisimple Lie group with Lie algebra $\mathcal{G}$. $\mathcal{G}$ has a natural Euclidean structure with the positive definite inner product given by the opposite of its Killing form. Now let us recall that $\mathfrak{D}(X, G)$ acts on $\mathfrak{D}_1(X, \mathcal{G})$ by the operators $V(g)$ such that, for all ω in $\mathfrak{D}_1(X, \mathcal{G})$, $V(g)\omega$ is the 1-form

$$x \to V(g)\omega(x) = \mathrm{Ad}\, g(x) \cdot \omega(x), \qquad x \in X$$

Taking into account the construction of energy scalar product $(\ ,\ )_F$ on $\mathfrak{D}_1(X, \mathcal{G})$ and the fact that the adjoint representation of G on $\mathcal{G}$ is unitary with respect to its natural Euclidean structure, it follows that V is unitary with respect to $(\ ,\ )_F$, and extends into a continuous unitary representation (also denoted V) of $\mathfrak{D}(X, G)$ on H_F.

Following the process described in Chapter 2, Section 3, we realize then a continuous unitary representation U_F of type (S) of $\mathfrak{D}(X, G)$ on the symmetric Fock space SH_F constructed with H_F as one particle space. U_F is defined by its action on the coherent states EXP ω, $\omega \in H_F$, by

$$\begin{aligned} U_F(g)\, \mathrm{EXP}\, \omega &= \exp[-\tfrac{1}{2}|b(g)|_F^2 - (V(g)\omega, b(g))_F] \\ &\quad \cdot \mathrm{EXP}(V(g)\omega + b(g)), \end{aligned}$$

where $g \in \mathfrak{D}(X, G)$, with $b(g) = dg \cdot g^{-1}$

Definition 3. U_F will be called the *basic energy representation* associated with the Riemannian flag F.

One easily sees that such a representation is of order 1 and that its support is the whole manifold X; moreover the corresponding spherical function φ_F, defined for all g in $\mathcal{D}(X, G)$ by $\varphi_F(g) = \langle U_f(g)\,\text{EXP}\,0, \text{EXP}\,0\rangle$ with respect to the vacuum state EXP 0 is such that

$$\varphi_F(g) = \exp[-\tfrac{1}{2}|dg \cdot g^{-1}|_F^2]$$

We want now to give another realization of U_F which has a useful and convenient form. Notice that φ_F is a continuous and positive definite function on the gauge group $\mathcal{D}(X, G)$, which induces on the nuclear space $\mathcal{D}_1(X, \mathcal{G})$ the positive definite (and continuous) function C_F such that

$$C_F(\omega) = \exp(-\tfrac{1}{2}|\omega|_F^2), \qquad \omega \in \mathcal{D}_1(X, \mathcal{G})$$

Therefore, from the Bochner-Minlos theorem, there exists a unique Gaussian measure μ_F with mean zero on the dual space $\mathcal{D}_1'(X, \mathcal{G})$ consisting of the $\mathcal{G}$-valued 1-currents on X, whose Fourier transform $\hat{\mu}_F$ is given by

$$\hat{\mu}_F(\omega) = \int_{\mathcal{D}_1'(X,\mathcal{G})} e^{i\langle\chi,\omega\rangle}\, d\mu_F(\chi) = C_F(\omega), \qquad \omega \in \mathcal{D}_1(X, \mathcal{G})$$

By Proposition 6, Chapter 2 (see also Theorem 7.9 of [9], see also e.g. [21]), we can realize U_F in the Hilbert space of L^2-Wiener functionals on $\mathcal{D}_1'(X, \mathcal{G})$, i.e., $L^2(\mathcal{D}_1'(X, \mathcal{G}); \mu_F)$.

In this picture U_F is such that for all g in $\mathcal{D}(X, G)$, all Φ in $L^2(\mathcal{D}_1'(X, \mathcal{G}); \mu_F)$, and for all χ in $\mathcal{D}_1'(X, \mathcal{G})$:

$$U_F(g)\Phi(\chi) = \exp[i\langle\chi, dg \cdot g^{-1}\rangle] \cdot \Phi(V^*(g^{-1})\chi)$$

where V^* denotes the extension of V obtained by transposition:

$$\langle V^*(g)\chi, \omega\rangle = \langle\chi, V(g^{-1})\omega\rangle$$

for all g in $\mathcal{D}(X, G)$, all χ in $\mathcal{D}_1'(X, \mathcal{G})$ and all ω in $\mathcal{D}_1(X, \mathcal{G})$.

The main problem is now to recognize what are the irreducible basic energy representations and to study the equivalence or nonequivalence of basic energy representations associated with different Riemannian flags.

3.3 THE REPRESENTATION $U_F^{\mathcal{A}}$ AND ITS SPECTRAL MEASURE

Let $F = (X_k, r_k)_{0\le k\le p}$ be a Riemannian flag of X, let U_F be the corresponding basic energy representation, and let $\mathcal{A}$ be a Cartan subalgebra of $\mathcal{G}$. The orthogonal complement of $\mathcal{A}$ in $\mathcal{G}$ (with respect to the natural scalar product of $\mathcal{G}$) will be denoted $\mathcal{A}^{\perp}$, and the maximal torus $\exp(\mathcal{A})$ in G will be denoted T.

We introduce the spaces $\mathcal{D}_1(X, \mathcal{A})$ and $\mathcal{D}_1(X, \mathcal{A}^{\perp})$ of respectively $\mathcal{A}$-valued and $\mathcal{A}^{\perp}$-valued compactly supported smooth 1-forms on X. From the orthogonal sum $\mathcal{G} = \mathcal{A} \oplus \mathcal{A}^{\perp}$ we get the orthogonal decomposition with respect to $(\ ,\ )_F$:

$$\mathcal{D}_1(X, \mathcal{G}) = \mathcal{D}_1(X, \mathcal{A}) \oplus \mathcal{D}_1(X, \mathcal{A}^{\perp})$$

The energy representation U_F defines a unitary representation $U_F^{\mathcal{A}}$ of the abelian nuclear group $\mathcal{D}(X, \mathcal{A}) = C_0^{\infty}(X, \mathcal{A})$ on the Hilbert space $L^2(\mathcal{D}_1'(X, \mathcal{G}); \mu_F)$ by

$$U_F^{\mathcal{A}}(u) = U_F(\exp u), \qquad u \in \mathcal{D}(X, \mathcal{A})$$

The present section is devoted to the study of $U_F^{\mathcal{A}}$.

Let d: $\mathcal{D}(X, \mathcal{A}) \to \mathcal{D}_1(X, \mathcal{A})$ be the exterior derivative, given by $u \to du$, where d is continuous with respect to the topologies of locally convex nuclear spaces of $\mathcal{D}(X, \mathcal{A})$ and $\mathcal{D}_1(X, \mathcal{A})$. Consequently the space of constant functions in $\mathcal{D}(X, \mathcal{A})$, i.e., ker d, is a closed subspace, which is equal to (0) if X is a noncompact manifold.

Let us denote by $\tilde{\mathcal{D}}(X, \mathcal{A})$ the space $\mathcal{D}(X, \mathcal{A})/\ker d$, and for any element u in $\mathcal{D}(X, \mathcal{A})$, by $\tilde{u}$ its class in $\tilde{\mathcal{D}}(X, \mathcal{A})$. $\tilde{\mathcal{D}}(X, \mathcal{A})$ is a nuclear space and the mapping $\tilde{d}$: $\tilde{\mathcal{D}}(X, \mathcal{A}) \to \mathcal{D}_1(X, \mathcal{A})$ defined by $\tilde{d}(\tilde{u}) = du$ is then a one-to-one continuous linear mapping. Consequently we can endow $\tilde{\mathcal{D}}(X, \mathcal{A})$ with a positive definite inner product $\tilde{F}$ by taking

$$\tilde{F}(\tilde{u}, \tilde{v}) = (\tilde{d}(\tilde{u}), \tilde{d}(\tilde{v}))_F = (du, dv)_F$$

We have easily the next result.

Lemma 1. $\tilde{d}$ is an isometry with close range from the real pre-hilbertian space $(\tilde{\mathcal{D}}(X, \mathcal{A}), \tilde{F})$ into the pre-hilbertian space $(\mathcal{D}_1(X, \mathcal{A}), (\ ,\ )_F)$. □

Let us consider the Gaussian measure $\tilde{\mu}_F$ on the dual space $\tilde{\mathcal{D}}'(X, \mathcal{A})$ with mean zero and Fourier transform $\hat{\tilde{\mu}}_F$ given by

$$\hat{\tilde{\mu}}_F(\tilde{u}) = \exp[-\tfrac{1}{2}\tilde{F}(\tilde{u}, \tilde{u})] = \exp[-\tfrac{1}{2}(du, du)_F]$$

Let us consider also the Gaussian measures m_F and $m_F^{\perp}$ on the spaces $\mathcal{D}_1'(X, \mathcal{A})$ and $\mathcal{D}_1'(X, \mathcal{A}^{\perp})$ given by their Fourier transforms:

$$\hat{m}_F\colon \omega \to \hat{m}_F(\omega) = \exp[-\tfrac{1}{2}(\omega, \omega)_F], \qquad \omega \in \mathcal{D}_1(X, \mathcal{A})$$

and

$$\hat{m}_F^{\perp}\colon \omega \to \hat{m}_F^{\perp}(\omega) = \exp[-\tfrac{1}{2}(\omega, \omega)_F], \qquad \omega \in \mathcal{D}_1(X, \mathcal{A}^{\perp})$$

Now, for all $\tilde{u}$ in $\tilde{\mathscr{D}}(X, \mathscr{A})$ let us define the operator $\tilde{W}(\tilde{u})$ on $L^2(\mathscr{D}_1'(X, \mathscr{A}); m_F)$ by

$$\tilde{W}(\tilde{u})\Phi(\chi) = \exp(i\langle\chi, du\rangle)\Phi(\chi)$$

for all Φ in $L^2(\mathscr{D}_1'(X, \mathscr{A}); m_F)$ and all χ in $\mathscr{D}_1'(X, \mathscr{A})$. Then

Lemma 2. (i) $\tilde{W}$: $\tilde{u} \to \tilde{W}(\tilde{u})$ is a continuous unitary representation of the nuclear abelian group $\tilde{\mathscr{D}}(X, \mathscr{A})$ on the Hilbert space $L^2(\mathscr{D}_1'(X, \mathscr{A}); m_F)$.

(ii) The spectral measure of $\tilde{W}$ is equivalent to $\tilde{u}_F$.

PROOF. (i) follows from an easy verification.

(ii) Let $\tilde{d}^*$ the transposed mapping of $\tilde{d}$; from Lemma 1 it follows that $\tilde{d}^*$ maps $\mathscr{D}_1'(X, \mathscr{A})$ onto $\tilde{\mathscr{D}}'(X, \mathscr{A})$. As these two spaces are standard Borel spaces there exists a Borel section s of $\tilde{d}^*$ such that the mapping

$$\tilde{s}: \chi \to (\tilde{d}^*\chi, s \cdot \tilde{d}^*\chi - \chi)$$

is an isomorphism of Borel spaces from $\mathscr{D}_1'(X, \mathscr{A})$ onto $\tilde{\mathscr{D}}'(X, \mathscr{A}) \times \ker(\tilde{d}^*)$. Consequently there exists a Borel measure λ on $\ker(\tilde{d}^*)$ such that

$$\tilde{s}(m_F) = \tilde{d}^*m_F \times \lambda$$

One then gets an isomorphism of Hilbert spaces:

$$L^2(\mathscr{D}_1'(X, \mathscr{A}); m_F) \simeq L^2(\tilde{\mathscr{D}}'(X, \mathscr{A}); \tilde{u}_F) \otimes L^2(\ker(\tilde{d}^*); \lambda)$$

which transforms, for all $\tilde{u}$ in $\tilde{\mathscr{D}}(X, \mathscr{A})$, the operator $\tilde{W}(\tilde{u})$ into the operator $\tilde{W}'(\tilde{u}) \otimes \mathbb{1}$, where $\tilde{W}'$ denotes the unitary representation of $\tilde{\mathscr{D}}(X, \mathscr{A})$ on $L^2(\tilde{\mathscr{D}}'(X, \mathscr{A}); \tilde{d}^*m_F)$ given for all $\tilde{u}$ in $\tilde{\mathscr{D}}(X, \mathscr{A})$ by

$$\tilde{W}'(\tilde{u})\Psi(\chi) = \exp[i\langle\chi, \tilde{d}(\tilde{u})\rangle] \cdot \Psi(\chi),$$

where $\Psi \in L^2(\tilde{\mathscr{D}}'(X, \mathscr{A}); \tilde{d}^*m_F)$, $\chi \in \tilde{\mathscr{D}}'(X, \mathscr{A})$

Consequently the spectral measure of $\tilde{W}$ is equivalent to $\tilde{d}^*m_F$; as all the spaces entering these relations are nuclear spaces and owing to the uniqueness of Gaussian measures given by their Fourier transforms, it follows that $\tilde{d}^*m_F$ equals $\tilde{\mu}_F$. So, the spectral measure of $\tilde{W}$ is equivalent to $\tilde{\mu}_F$. □

Let R^F be the unitary representation of $\mathscr{D}(X, \mathscr{A})$ on $L^2(\mathscr{D}_1'(X, \mathscr{A}^\perp); m_F^\perp)$ given by

$$R^F(u)\Phi(\chi) = \Phi(V^*(\exp u)\chi), \qquad u \in \mathscr{D}(X, \mathscr{A})$$

for all Φ in $L^2(\mathscr{D}_1'(X, \mathscr{A}^\perp); m_F^\perp)$, χ in $\mathscr{D}_1'(X, \mathscr{A}^\perp)$.

Lemma 3. The spectral measure ν^F of R^F is equivalent to the infinite direct sum $\bigoplus_{k\geq 0}(dv_F \otimes N)^{\otimes k}$, where dv_F is the measure on X such that, for all f in $C_0^\infty(X)$,

$$\int_X f(x)\, dv_F(x) = \sum_{k=0}^{p} \int_{X_k} f(x_k)\, dv_k(x_k)$$

and N being the counting measure on the set Δ of roots of $\mathcal{G}$ with respect to the Cartan subalgebra $\mathcal{A}$.

PROOF. We shall use the symmetric Fock realization of $L^2(\mathcal{D}_1'(X, \mathcal{A}^\perp); m_F^\perp)$: let $E(F)$ be the complex Hilbert space spanned by $\mathcal{D}_1(X, \mathcal{A}^\perp)$ with respect to the scalar product $(\ ,\ )_F$; by proposition 3, Chapter 2, one knows that $L^2(\mathcal{D}_1'(X, \mathcal{A}^\perp); m_F^\perp)$ is isomorphic to the symmetric Fock space:

$$SE(F) = \sum_{k\geq 0}^{\oplus} S^k E(F)$$

where $S^kE(F)$ denotes the kth symmetric tensor power of $E(F)$. Let Δ be the set of roots of the symmetric pair $(\mathcal{G}, \mathcal{A})$; to each α in Δ corresponds the subspace $\mathcal{G}^\alpha$ of $\mathcal{G}$ with weight α. From the decomposition $\mathcal{A}^\perp = \bigoplus_{\alpha\in\Delta}\mathcal{G}^\alpha$, it follows that we have the decomposition $E(F) = \bigoplus_{\alpha\in\Delta}E_\alpha$, E_α denoting the complex Hilbert space spanned by $\mathcal{D}_1(X, \mathcal{G}^\alpha)$ with respect to $(\ ,\ )_F$. Consequently we have the tensor decomposition

$$SE(F) = \bigotimes_{\alpha\in\Delta} SE_\alpha$$

Notice that for all ω in E_α and all u in $\mathcal{D}(X, \mathcal{A})$ one has $V(\exp u)\omega(x) = \exp[i\langle\alpha, u(x)\rangle]\omega(x)$, $x \in X$. The restriction of R_F to $S^kE(F)$ (which equals the space $\otimes_{\alpha\in\Delta}S^kE_\alpha$) acts by multiplication by elements of the form $\exp[i \sum_{s=1}^k \alpha_s(u(x_s))]$, with $\alpha_1, \ldots, \alpha_k$ in Δ and $x_1, \ldots, x_k$ in X.

It follows that the spectral measure ν^F of R^F is supported by the subset Γ of $\mathcal{D}'(X, \mathcal{A})$ consisting of functionals of the form

$$\chi\colon u \to \chi(u) = \sum_{s=1}^{k} \alpha_s(u(x_s))$$

for all nonnegative integers k, the α_s being in Δ, and the x_s in X. We can then identify Γ with $\cup_{k=0}^\infty(X \times \Delta)^k_{\text{sym}}$, and ν^F with the Poisson measure whose restriction to the kth symmetric power $(X \times \Delta)^k_{\text{sym}}$ is given by $(dv_F \otimes N)^{\otimes k}$. Here N is the counting measure on Δ and dv_F is the measure on X coming from the Riemannian flag of X (see, e.g., [18, Section 3]). □

Let us come back to the representation $U_F^{\mathcal{A}}$ of $\mathcal{D}(X, \mathcal{A})$ on $L^2(\mathcal{D}_1'(X, \mathcal{G}); \mu_F)$ given by $U_F^{\mathcal{A}}(u) = U_F(\exp u)$. From the isomorphism

$$L^2(\mathcal{D}_1'(X, \mathcal{G}); \mu_F) \simeq L^2(\mathcal{D}_1'(X, \mathcal{A}); m_F) \otimes L^2(\mathcal{D}_1'(X, \mathcal{A}^\perp); m_F^\perp)$$

it follows easily that, for all u in $\mathcal{D}(X, \mathcal{A})$

$$U_F^{\mathcal{A}}(u) = \tilde{w}(\tilde{u}) \otimes R^F(u)$$

From Lemmas 2 and 3, one gets then

Theorem 1. Let $F = (X_k, r_k)_{0\le k\le p}$ be a Riemannian flag of X, let U_F be the corresponding energy representation of the gauge group $\mathfrak{D}(X, G)$, let $\mathcal{A}$ be a Cartan subalgebra of $\mathcal{G}$, and let $U_F^{\mathcal{A}}$ be the associated representation of $\mathfrak{D}(X, \mathcal{A})$. The spectral measure of $U_F^{\mathcal{A}}$ is equivalent to the convolution $\tilde{u} * \nu^F$, where $\tilde{\mu}_F$ is the Gaussian measure on $\tilde{\mathfrak{D}}'(X, \mathcal{A})$ with Fourier transform $\tilde{u} \to \exp[-|du|_F^2/2]$ and with $\omega^F \simeq \bigoplus_{k\ge 0}(dv_F \otimes N)^{\otimes k}$. □

We want to give now a direct integral decomposition of $U_F^{\mathcal{A}}$ · Let χ be in the subset Γ of $\mathfrak{D}'(X, \mathcal{A})$; then $\chi = \Sigma_{j\in J}\delta_{x_j}^{\alpha_j}$ where J is any finite subset of I, the α_j are in Δ and the x_j in X, $\delta_{x_j}^{\alpha_j}$ being the functional such that for all u in $\mathfrak{D}(X, \mathcal{A})$, $\delta_{x_j}^{\alpha_j}(u) = \alpha_j(u(x_j))$. We define a character γ_χ of $\mathfrak{D}(X, \mathcal{A})$ by

$$\gamma_\chi(u) = \exp(i\langle\chi, u\rangle)$$

Let us consider now $\tilde{W}_\chi = \tilde{W} \otimes \gamma_\chi$, $\tilde{W}$ being the unitary representation defined in Lemma 2. Let W be the unitary representation of $\mathfrak{D}(X, \mathcal{A})$ defined by $W(u) = \tilde{W}(\tilde{u})$; W and $\tilde{W}$ have the same spectral measure, equivalent to $\tilde{\mu}_F$ by Lemma 2. It follows that the spectral measure $\tilde{\mu}_F^\chi$ of $\tilde{W}_\chi$ is the convolution of the spectral measure $\tilde{\mu}_F$ of W by the spectral measure of the character γ_χ.

Lemma 4. (i) One has the direct integral decomposition

$$U_F^{\mathcal{A}} = \int_\Gamma^\oplus \tilde{W}_\chi \cdot d\nu^F(\chi)$$

(ii) $\tilde{\mu}_F^\chi$ is equivalent to the translated of $\tilde{\mu}_F$ by χ, i.e., $\tilde{\mu}_F^\chi \simeq \tilde{\mu}_F(\cdot - \chi)$.

PROOF. (i) From Lemma 3 it follows that $R^F = \int_\Gamma^\oplus \gamma_\chi \cdot d\nu^F(\chi)$, and then

$$U_F^{\mathcal{A}} = \tilde{W} \otimes R^F = \tilde{W} \otimes \int_\Gamma^\oplus \gamma_\chi \cdot d\nu^F(\chi) = \int_\Gamma^\oplus (W \otimes \gamma_\chi)\, d\nu^F(\chi)$$

$$= \int_\Gamma^\oplus \tilde{W}_\chi\, d\nu^F(\chi)$$

(ii) follows from the fact that the spectral measure of $\tilde{W}_\chi$ is the convolution of the spectral measure $\tilde{\mu}_F$ of $\tilde{W}$ by the spectral measure of the character γ_χ which is given by χ, and then: $\tilde{\mu}_F^\chi = \tilde{\mu}_F(\cdot - \chi)$. □

REMARKS. (1) Let $\mathcal{U}$ be the algebra of measurable subsets of $\mathfrak{D}'(X, \mathcal{A})$; the mapping $(\chi, B) \in \Gamma \times \mathcal{U} \to \tilde{\mu}_F^\chi(\chi, B)$ is measurable because $\tilde{\mu}_F^\chi(\chi, B) = \tilde{\mu}_F(B - \chi)$.

(2) The results of Lemmas 2, 3, 4 and Theorem 1 were provided in [7] and [23] in the case of a noncompact manifold X and with F reduced

to the case $F = (X = X_0, r_0)$, i.e., with X endowed with a smooth Riemannian structure.

3.4 Γ-PROPERTY; Γ′-PROPERTY

Let $\mathcal{A}$ be a Cartan subalgebra of $\mathcal{G}$. For a Gaussian measure μ on $\tilde{\mathcal{D}}'(X, \mathcal{A})$ and a Borel measure ν on Γ, $\mu * \nu$ is the measure on $\tilde{\mathcal{D}}'(X, \mathcal{A})$ such that if A is any Borel subset of $\tilde{\mathcal{D}}'(X, \mathcal{A})$ then

$$(\mu * \nu)(A) = \int_\Gamma \mu(A + \chi)\, d\nu(\chi)$$

Definition 4. (1) A Gaussian measure μ on $\tilde{\mathcal{D}}'(X, \mathcal{A})$ will be said to have the Γ-property if for ν_1 and ν_2 mutually singular probability measures on Γ, then $\mu * \nu_1$ and $\mu * \nu_2$ are singular with respect to each other.

(2) A pair (μ^1, μ^2) of Gaussian measures on $\mathcal{D}'(X, \mathcal{A})$ will be said to have the Γ′-property if for any Borel measure ν on Γ with $\nu(\{0\}) = 0$, then $\mu^1 * \nu$ and $\mu^2 * \nu$ are singular with respect to each other.

Let us recall that to each Riemannian flag F of X there is associated a Gaussian measure $\tilde{\mu}_F$ with mean zero on $\tilde{\mathcal{D}}'(X, \mathcal{A})$, whose Fourier transform is given by $\tilde{u} \to \exp[-(1/2)(du, du)_F]$.

We give now two important results which constitute a generalization of the results given in [7] for the case where Riemannian flags are reduced to single Riemannian manifolds.

Theorem 2. Let F be a Riemannian flag of X. $\tilde{\mu}_F$ has the Γ-property in the following two cases:

(1) $d_0 = \dim(X) \geq 3$;
(2) $d_0 = 2$ and, moreover, for any root λ of the pair $(\mathcal{G}, \mathcal{A})$, its length $|\lambda|$ satisfies $|\lambda| \geq K_F$, where K_F is a positive constant depending only on F.

Theorem 3. Let F^1 and F^2 be two different Riemannian flags of X. The pair $(\tilde{\mu}_{F^1}, \tilde{\mu}_{F^2})$ has the Γ′-property in the following two cases:

(1) $d_0 \geq 3$;
(2) $d_0 = 2$ and, moreover, for any root λ of the pair $(\mathcal{G}, \mathcal{A})$ one has $|\lambda| \geq \max(K_{F^1}, K_{F^2})$.

For the proof of these results we refer to [18]. □

REMARKS. (1) Let F be a Riemannian flag of X, and let $\mathcal{A}_1$ and $\mathcal{A}_2$ be two Cartan subalgebras of $\mathcal{G}$. Let us denote by Δ_i the corresponding root system, and by Γ_i the set $\cup_{k\geq 0}(X \times \Delta_i)^k_{\text{sym}}$ endowed with the Poisson measure ν_i^F, $i = 1, 2$. From the isomorphism of the Borel measured spaces

(Γ_1, ν_1^F) and (Γ_2, ν_2^F) one easily concludes that the Γ-property and Γ'-property do not depend on a particular choice of Cartan subalgebra of $\mathcal{G}$.

(2) As $\tilde{\mu}_F^\chi$ is equivalent to the spectral measure of $\tilde{W}_\chi$, χ in Γ, the Γ-property implies that for any pair (A_1, A_2) of Borel subsets of Γ such that $\nu^F(A_1) > 0$, $\nu^F(A_2) > 0$, $\nu^F(A_1 \cap A_2) = 0$, the representations $\int_{A_1}^{\oplus} \tilde{W}_\chi \, d\nu^F(\chi)$ and $\int_{A_2}^{\oplus} \tilde{W}_\chi \, d\nu^F(\chi)$ are disjoint, i.e., do not contain equivalent subrepresentations.

An important consequence of the Γ-property is

Lemma 5. Let F be a Riemannian flag of X, $\mathcal{A}$ a Cartan subalgebra of $\mathcal{G}$, and let us assume that $\tilde{\mu}_F$ has the Γ-property. The von Neumann algebra generated by $U_F^{\mathcal{A}}$ contains all the operators of the form $\mathbb{1} \otimes R^F(u)$ and of the form $W(u) \otimes \mathbb{1}$, with u in $\mathcal{D}(X, \mathcal{A})$.

PROOF. Let us recall that $W(u) = \tilde{W}(\tilde{u})$, $\tilde{u} \in \tilde{\mathcal{D}}(X, \mathcal{A})$. By Lemma 4 one has

$$U_F^{\mathcal{A}} = \int_\Gamma^{\oplus} \tilde{W}_\chi \cdot d\nu^F(\chi)$$

Thus $U_F^{\mathcal{A}}$ is equivalent to a representation of $\mathcal{D}(X, \mathcal{A})$ on a direct integral of Hilbert spaces $\int_\Gamma^{\oplus} H_\chi \, d\nu^F(\chi)$. Let S be an operator commuting with $U_F^{\mathcal{A}}$, and let Γ' be a measurable subset of Γ satisfying $\nu^F(\Gamma') > 0$ and $\nu^F(\Gamma - \Gamma') > 0$; from the Γ-property it follows that $\int_{\Gamma'}^{\oplus} H_\chi \cdot d\nu^F(\chi)$ and its orthogonal complement $\int_{\Gamma-\Gamma'}^{\oplus} H_\chi \cdot d\nu^F(\chi)$ do not contain equivalent subrepresentations of $U_F^{\mathcal{A}}$, although they are invariant by S. It follows that S is decomposable into a direct integral with respect to ν^F.

Let $\mathrm{VN}(F)$ be the von Neumann algebra generated by $U_F^{\mathcal{A}}$. The operators $(\mathbb{1} \otimes R^F)(u)$ and $(W \otimes \mathbb{1})(u)$ act on each Hilbert space H_χ by multiplication by a bounded ν^F-measurable function; consequently they commute with all the decomposable operators, in particular with operators S which commute with $U_F^{\mathcal{A}}$; these operators are then in the bicommutant of $U_F^{\mathcal{A}}$, and thus in $\mathrm{VN}(F)$. □

As a corollary one gets

Corollary. $\mathrm{VN}(F)$ contains all operators of multiplication by functions of the form $\exp[i(\cdot, du)_F]$, $u \in \mathcal{D}(X, \mathcal{A})$. □

3.5 ON THE IRREDUCIBILITY OF THE BASIC ENERGY REPRESENTATION U_F

Theorem 4. Let F be a Riemannian flag of X. If $d_0 = \dim(X) \geq 3$, or if $d_0 = 2$ with the additional condition that $|\lambda| \geq K_F$ for all roots of $\mathcal{G}$, then the basic energy representation U_F is irreducible.

PROOF. We shall use the realization of U_F in the space of L^2-Wiener functionals $L^2(\mathcal{D}_1'(X, \mathcal{G}); \mu_F)$ given in Section 3.2.

Let us prove, at first, that the "vacuum vector" **1**, $\chi \to \mathbf{1}(\chi) = 1$, $\forall\, \chi \in \mathcal{D}_1'(X, \mathcal{G})$, is cyclic for U_F. Let $L(F)$ be the von Neumann algebra generated by U_F; as $U_F^{\mathcal{A}}(u) = U_F(\exp u)$, $u \in \mathcal{D}(X, \mathcal{A})$, the bicommutant of U_F contains the bicommutant of $U_F^{\mathcal{A}}$. From the assumption made in Theorem 4, it follows from Theorem 2 that $\tilde{\mu}_F$ has the Γ-property; consequently by the corollary of Lemma 5, $L(F)$ contains the operators of multiplication by $\exp[i(\cdot, du)_F]$, $u \in \mathcal{D}(X, \mathcal{A})$, for any Cartan subalgebra $\mathcal{A}$ of $\mathcal{G}$. As $\mathcal{G}$ is the union of all its Cartan subalgebras, it follows that $L(F)$ contains all operators of multiplication by $\exp[i(\cdot, du)_F]$ for all u in $\mathcal{D}(X, \mathcal{G})$, and thus all the operators of the form

$$\eta_F^u(g) = U_F(g) \cdot \exp[i(\cdot, du)_F] \cdot U_F(g^{-1})$$

for all g in $\mathcal{D}(X, G)$ and all u in $\mathcal{D}(X, \mathcal{G})$. Consequently $L(F)$ contains all the operators of multiplication by

$$\exp[i(\cdot, V^*(g)\, du)_F], \qquad g \in \mathcal{D}(X, \mathcal{G}),\ u \in \mathcal{D}(X, \mathcal{G}).$$

It follows that $L(F)$ contains all the operators of multiplication by functions of the form $\exp[i(\cdot, \Sigma_{k=1}^m V^*(g_k)\, du_k)_F]$, for all finite subsets $\{g_1, \ldots, g_m\}$ in $\mathcal{D}(X, G)$ and $\{u_1, \ldots, u_m\}$ in $\mathcal{D}(X, \mathcal{G})$.

Moreover, by Lemma 3.5 of [7] one knows that the set $\{V^*(g)\, du | (g, u) \in \mathcal{D}(X, G) \times \mathcal{D}(X, \mathcal{G})\}$ is total in $\mathcal{D}_1(X, \mathcal{G})$. It follows that the functions $\chi \to \eta_f^u(g)(\chi)$, $(g, u) \in \mathcal{D}(X, G) \times \mathcal{D}(X, \mathcal{G})$ form a total set in $L^2(\mathcal{D}_1'(X, \mathcal{G}); \mu_F)$.

As $U_F(g)\mathbf{1} = \exp[i(\cdot, dg \cdot g^{-1})_F]$, it follows that the smallest closed subspace of $L^2(\mathcal{D}_1'(X, \mathcal{G}); \mu_F)$ containing $U_F(g)\mathbf{1}$, $g \in \mathcal{D}(X, G)$, contains the space spanned by the functions $\eta_F^u(g)$, and thus the whole space $L^2(\mathcal{D}_1'(X, \mathcal{G}); \mu_F)$. The cyclicity of **1** is then proved.

There remains now to prove the irreducibility of the cyclic component of U_F (with respect to **1**). Let Q be in the commutant of $U_F(\mathcal{D}(X, G))$; in particular, Q commutes with $U_F^{\mathcal{A}}$ for any Cartan subalgebra $\mathcal{A}$ of $\mathcal{G}$. As shown in the proof of Lemma 5, this implies that Q is decomposable with respect to the integral decomposition of $U_F^{\mathcal{A}}$ given in Lemma 4, for any Cartan subalgebra $\mathcal{A}$ of $\mathcal{G}$. The projection of $L^2(\mathcal{D}_1'(X, \mathcal{G}); \mu_F)$ onto $L^2(\mathcal{D}_1'(X, \mathcal{A}); m_F)$ being diagonalizable, the space $L^2(\mathcal{D}_1'(X, \mathcal{A}); m_F)$ is invariant by Q. It follows that $Q\ \mathbf{1}$ belongs to all the spaces $L^2(\mathcal{D}_1'(X, \mathcal{A}); m_F)$ for all Cartan subalgebras $\mathcal{A}$ of $\mathcal{G}$, whose intersection equals $\mathbb{C}\ \mathbf{1}$. Consequently, Q is a scalar operator, and then U_F is irreducible. □

REMARKS. (1) In the next chapter we shall see that in the case $\dim(X) = d_0 = 1$ (for which F is necessarily reduced to (X, r) where r is a C^1-Riemannian structure on the one dimensional manifold X), the basic energy representation U_F is always reducible.

(2) An a priori possible generalization of the definition of the inner product (, $)_F$ is obtained by considering

$$(\omega, \omega')_F = \sum_{k=0}^{p} \int_{X_R} \mathrm{tr}(\omega_k'(x)^* \ \omega_k(x)) p_k(x) \ dv_k(x)$$

where p_k is a strictly positive function on X_k. Actually, except in the case $\dim X_k = 2$, this generalization has no interest since one can always incorporate the weight p_k in the volume measure, just using a local dilation of the Riemannian structure. Nevertheless, the irreducibility result we proved in this section depends on the choice of the weight $pd_0 - 2$ even if one chooses this weight to be a constant.

All new facts coming from the introduction of weights are concentrated on the two-dimensional manifold. For the sake of simplicity, let us examine the case of a trivial Riemannian flag of (maximal) dimension 2 ($k = 0$, $d_0 = 2$ in the notations of Definition 3.1). The only modification consists in a new definition for the scalar product

$$(\omega, \omega')_F = \int_X \mathrm{tr}(\omega'(x)^* \ \omega(x)) p(x) \ dv(x)$$

in which we dropped all k-indices since they only take the value 0. The corresponding energy representation will be said to be weighted.

In this context, one can be much more precise about the constant K_F appearing in the main theorem 4. Optimal results in this situation can be stated as follows.

Theorem 5. If $\dim X = 2$, the weighted energy representation is irreducible if any root α of g satisfies

$$\|\alpha\| > \sup_{x \in X} \left(\frac{8\pi}{p(x)} \right)^{1/2} \equiv \alpha_0 \qquad (*)$$

where, in the first member, the norm of α is evaluated with the positive scalar product on $\mathcal{G}$ given by the opposite of the Killing form.

Theorem 6. Assuming that two weighted energy representations on X, with $\dim X = 2$, satisfy (*), then these representations are disjoint except if the corresponding scalar products are identical.

These results were conjectured in [7], on the basis of the relation with the problem of the triviality of the two-dimensional exponential interaction model (Høegh-Krohn or $:e^{\beta\varphi}:_2$-model) in constructive quantum field theory. It was namely showed that the $:e^{\beta\varphi}:_2$-model (over $\mathbb{R}^2$) is trivial, in a very precise sense, when $|\beta| > \sqrt{8\pi}$ [1, 2, 4, 14, 19, 20] and is not trivial for $|\beta| < \sqrt{4\pi}$ in the L^2-sense [2] and for $|\beta| < \sqrt{8\pi}$ in the $L^{1+\varepsilon}$-sense [14]. The value $\sqrt{8\pi}$ for β corresponds to the value α_0 for α.

In [7] a weaker result with 8 replaced by 32 in α_0 was proved. After a careful analysis of the disjointness of Gaussian measures with Laplacian-like covariances in dimension 2 the proofs of Theorems 6 and 7 was obtained by Wallach in [24]. Up to now, the case where the condition (*) is not fulfilled is completely open. More precisely, the result of Wallach and the analysis in [7] show that the essential ingredient of the proof, as it stands, breaks down for $\|\alpha\|$ strictly less than α_0. The recent result [11, 20] on triviality of the $:e^{\beta\varphi}:_2$-interaction for $\beta = \sqrt{8\pi}$ indicates that irreducibility of the energy representation could also be expected for $\|\alpha\| = \alpha_0$. To handle irreducibility or reducibility for $\|\alpha\| < \alpha_0$ it seems that a new method is needed. Such a method could also give new tools for the mathematical construction of a quantized (Euclidean) nonlinear σ-model (because of the above-mentioned relation of representation theory with constructive quantum field theory and the relation with the one-dimensional case discussed in Chapter 4). Let us finally remark that the values $|\beta| = \sqrt{8\pi}$ (resp. $|\beta| = \sqrt{4\pi}$) corresponds to dimension 19 (resp. 13) for a bosonic (Polyakov) string model; cf. [6, 8].

Roughly speaking Theorem 4 proves the Γ-property implies the irreducibility; at the present time we do not know whether the converse is true.

Using the Γ'-property, by the same arguments as the ones used in Theorem 4, one easily proves

Theorem 7. Let F and F' be two different Riemannian flags of X. Under the assumption of Theorem 3, the basic energy representation U_F and $U_{F'}$ are not unitarily equivalent.

Notice that if $\dim(X) \geq 3$, the assignment $F \to U_F$ defines a one-to-one mapping form the set of Riemannian flags of X into the unitary dual of $D(X; G)$, and, more precisely, into the set of nonlocated and order 1 G-distribution on X.

3.6 RINGS OF GENERALIZED ENERGY REPRESENTATIONS

3.6.1. The Generalized Energy Representation $U_F^{(n,p)}$

Let F be a Riemannian flag of the manifold X; the associated energy Gaussian measure μ_F on $\mathcal{D}_1'(X, \mathcal{G})$ is invariant by the operators $V^*(g)$, $g \in$

$\mathcal{D}(X, G)$, because of the unitarity of the operators $V(g)$. In other words, for all g in $\mathcal{D}(X, G)$ one has $\mu_F \circ V^*(g) = \mu_F$. It follows that for each positive integer n, the product measure $\mu_F^n = \mu_F \times \cdots \times \mu_F$ (n copies) on the product space $\mathcal{D}_1'(X, \mathcal{G})^n$ is invariant by the operators $V^*(g) \times \cdots \times V^*(g)$. On the other side, when we look at the form of the basic energy representation U_F we see that we can write, for all g in $\mathcal{D}(X, G)$,

$$U_F(g) = W_g \cdot R^F(g)$$

where R^F is the regular representation of $\mathcal{D}(X, G)$ on $L^2(\mathcal{D}_1'(X, G); \mu_F)$ given, for all Φ in $L^2(\mathcal{D}_1'(X, G); \mu_F)$ by

$$R^F(g)\,\Phi\,(\chi) = \Phi(V^*(g^{-1})\chi),$$

which leaves invariant the measure $\mu_F \circ W_g$ is the operator of multiplication by $\exp[i\langle \cdot, dg \cdot g^{-1}\rangle]$:

$$W_g\,\Phi(\chi) = \exp[i\langle \chi, dg \cdot g^{-1}\rangle]\,\Phi(x)$$

Thus, although μ_F is invariant by V^* (and not merely quasi-invariant), we are in a case which is formally similar to the one that allows the Weyl construction of finite-dimensional representations of the general linear group, and of finite functional dimensional representations of the group of diffeomorphisms of a noncompact manifold [12]. From this remark one of us [17] constructed new nonlocated and order 1 unitary representations of $\mathcal{D}(X, G)$, called *generalized energy representations.* These representations include as a particular case the basic energy representations.

So, let us select a Riemannian flag F of X, let n be a positive integer, and let ρ be a unitary representation of the symmetric group S_n on some Hilbert space E_ρ. We associate with these data the Hilbert space $L_F^2(n, \rho)$ of functionals Φ defined on $\mathcal{D}_1'(X, \mathcal{G})^n$ and with values in E_ρ such that

$$\|\Phi\|_{F,n,\rho}^2 = \int_{\mathcal{D}_1'(X,\mathcal{G})^n} \|\Phi\,(\chi_1, \ldots, \chi_n)\|_{E_\rho}^2\, d\mu_F^n(\chi_1, \ldots, \chi_n) < \infty$$

Now, for all g in $\mathcal{D}(X, G)$, let us consider the operators $U_{F,n}(g)$ defined for all Φ in $L_F^2(n, \rho)$ by

$$\begin{aligned} U_{F,n}(g)\,\Phi(\chi_1, \ldots, \chi_n) &= \exp\left[i\left\langle \sum_{k=1}^{n} \chi_k, dg \cdot g^{-1}\right\rangle\right] \\ &\quad \cdot \Phi(V^*(g^{-1})\,\chi_1, \ldots, V^*(g^{-1})\chi_n) \\ &= \prod_{k=1}^{n} \exp[i\langle \chi_k, dg \cdot g^{-1}\rangle] \\ &\quad \cdot \Phi(V^*(g^{-1})\,\chi_1, \ldots, V^*(g^{-1})\chi_n), \end{aligned}$$

where $(\chi_1, \ldots, \chi_n) \in \mathcal{D}'_1(X, \mathcal{G})^n$

From the above discussion the next Lemma follows.

Lemma 6. The operators $U_{F,n}(g)$, $g \in \mathcal{D}(X, G)$, are unitary operators on $L_F^2(n, \rho)$ and the mapping $g \to U_{F,n}(g)$ is a continuous unitary representation of $\mathcal{D}(X, G)$ on $L_F^2(n, \rho)$, which is nonlocated and of order 1.

REMARKS. (1) For $n = 1$ and ρ the unit representation of S_1 on $\mathbb{C}$, $U_{F,1}$ is exactly the basic energy representation U_F.

(2) Let u be a unitary vector in E_ρ, and let $\mathbf{u}$ be the constant element in $L^2(\mathcal{D}'_1(X, \mathcal{G})^n; \mu_F^n)$ which equals u everywhere. The spherical function $g \to (U_{F,n}(g)\mathbf{u}, \mathbf{u})_{F,n,\rho}$ is equal to

$$\int_{\mathcal{D}'_1(X,\mathcal{G})^n} \prod_{k=1}^{n} \exp[i\langle \chi_k, dg \cdot g^{-1}\rangle]\, d\mu_F(\chi_1) \cdots d\mu_F(\chi_n)$$

$$= \prod_{k=1}^{n} \int_{\mathcal{D}'_1(X,\mathcal{G})^n} \exp[i\langle \chi_k, dg \cdot g^{-1}\rangle]\, d\mu_F(\chi_k)$$

$$= [\exp(-\tfrac{1}{2}|dg \cdot g^{-1}|_F^2)]^n$$

It follows that $U_{F,n}$ is always reducible for $n > 1$ and appears as a tensor product of n copies of the basic energy representation U_F.

Let us consider the closed subspace $H_F^{(n,\rho)}$ of $L_F^2(n, \rho)$ consisting of functionals Φ such that, for σ in the symmetric group S_n and all $(\chi_1, \ldots, \chi_n)$ in $\mathcal{D}'_1(X, \mathcal{G})^n$,

$$\Phi(\chi_{\sigma(1)}, \ldots, \chi_{\sigma(n)}) = \rho^{-1}(\sigma)\Phi(\chi_1, \ldots, \chi_n)$$

We have

Theorem 8. The restriction $U_F^{(n,\rho)}$ of $U_{F,n}$ to $H_F^{(n,\rho)}$ is a nonlocated unitary representation of order 1 of $\mathcal{D}(X, G)$.

PROOF. An easy computation shows that the subspace $H_F^{(n,\rho)}$ is invariant by all the operators $U_{F,n}(g)$; the assertion follows then from Lemma 6.

REMARK. If ρ_1 and ρ_2 are two unitarily equivalent representations of S_n, then $U_F^{(n,\rho_1)}$ and $U_F^{(n,\rho_2)}$ are unitarily equivalent.

3.6.2. The Ring $A_F(X, G)$

Let us consider the fibered set $\hat{S} = \cup_{n\geq 1}\hat{S}_n$ consisting of pairs (n, ρ) where n is a positive integer and ρ a (class of) unitary representation of S_n ($\rho \in \hat{S}_n$). We define two operations on $\tilde{S}$. The first is defined on each "fiber"

$\hat{S}_n$ of $\hat{S}$ and corresponds to the direct orthogonal sum: let n be a positive integer, and let ρ_1, ρ_2 be in $\hat{S}_n$; then we define

$$(n, \rho_1) \oplus (n, \rho_2) = (n, \rho_1 \oplus \rho_2)$$

The second law, known as the exterior product [13], is defined as follows: let (n_1, ρ_1) and (n_2, ρ_2) be in $\hat{S}$, and let us consider S_{n_1} and S_{n_2} as groups of permutations of $\{1, \ldots, n_1\}$ and $\{n_1 + 1, \ldots, n_1 + n_2\}$ respectively; we have then a natural embedding of $S_{n_1} \times S_{n_2}$ as a subgroup of $S_{n_1+n_2}$. The exterior product, denoted $\circ$, is defined by

$$(n_1, \rho_1) \circ (n_2, \rho_2) = (n_1 + n_2, \underset{S_{n1} \times S_{n2} \uparrow S_{n1+n2}}{\mathrm{Ind}} \rho_1 \times \rho_2)$$

In [17] the following theorem is proved.

Theorem 9. Let F be a Riemannian flag of X, and let $A_F(X, G)$ be the set of (classes of) generalized energy representations $U_F^{(n,\rho)}$, $(n, \rho) \in \hat{S}$. Then

(1) For all $n \geq 1$, all ρ_1, ρ_2 in $\tilde{S}_n$ the representations $U_F^{(n,\rho_1)} \oplus U_F^{(n,\rho_2)}$ and $U_F^{(n,\rho_1 \oplus \rho_2)}$ are unitarily equivalent.

(2) For all (n_1, ρ_1) and (n_2, ρ_2) in $\tilde{S}$, the representations $U_F^{(n_1,\rho_1)\circ(n_2,\rho_2)}$ and $U_F^{(n_1,\rho_1)} \oplus U_F^{(n_2,\rho_2)}$ are unitarily equivalent.

As a corollary one gets

Corollary. $A_F(X, G)$ is closed under direct sum and tensor product.

REMARKS. (1) $U_F = U_F^{(1,\mathbf{1})}$ where **1** denotes here the unit representation of S_1; in particular U_F belongs to $A_F(X, G)$.

(2) From assertion (1) of Theorem 9, it follows that if $\rho \in \hat{S}_n$ is reducible, then $U_F^{(n,\rho)}$ is reducible.

REFERENCES

1. S. Albeverio, G. Gallavotti, and R. Høegh-Krohn, *Comm. Math. Phys.*, *70*:187 (1979).
2. S. Albeverio and R. Høegh-Krohn, *J. Funct. Anal.*, *16*:39 (1974).
3. S. Albeverio and R. Høegh-Krohn, *Comp. Math.*, *36*:37 (1978).
4. S. Albeverio and R. Høegh-Krohn, *Quantum Fields-Algebras, Processes* (L. Streit, ed.), Springer-Verlag, Wien, pp. 331–335, 1980.
5. S. Albeverio, R. Høegh-Krohn, and H. Holden, *J. Funct. Anal.*, *78*:154 (1986).
6. S. Albeverio, R. Høegh-Krohn, S. Paycha, and S. Scarlatti, *Acta Appl. Math.* *26*:103 (1992).
7. S. Albeverio, R. Høegh-Krohn, and D. Testard, *J. Funct. Anal.*, *41*:378 (1981).

8. S. Albeverio, J. Jost, S. Paycha, and S. Scarlatti, *A Mathematical Introduction to String Theory*, monograph submitted for publication.
9. A. Guichardet, *Symmetric Hilbert Spaces and Related Topics*, Lecture Notes in Math., vol. 26, Springer-Verlag, Berlin-Heidelberg, New York (1972).
10. R. Ismagilov, *Mat. U.S.S.R. Sb. 29*:117 (1976); (in Russian; transl.) *29*:105 (1976).
11. J. P. Kahane, *Ann. Sci. Math. Québec*, *9*:105 (1985).
12. A. A. Kirillov, *Sel. Math. Sov.*, *1*:351 (1981).
13. D. Knutson, *λ-rings and the representations theory of the symmetric group*, Lecture Notes in Math., vol. 308, Springer-Verlag, Berlin, Heidelberg, New York (1973).
14. S. Kusuoka, *Ideas and Methods in Quantum and Statistical Physics*, (S. Albeverio, J. E. Fenstad, H. Holden, T. Lindstrøm, eds.), Cambridge University Press, pp. 405–424 (1992).
15. J. Marion, *J. Funct. Anal.*, *54*:1 (1983).
16. J. Marion, *Publi. Mat. 34*:3 (1990).
17. J. Marion, *Publi. Mat.*, *33*:99 (1989).
18. J. Marion and D. Testard, *J. Funct. Anal.*, *76*:160 (1988).
19. Y. Osipov, *Rep. Math. Phys. 20*:111 (1984).
20. H. Sato, Kyushu University manuscript (1990).
21. I. Segal: *J. Funct. Anal.*, *33*:175 (1979).
22. A. M. Vershik, I. M. Gelfand, and M. I. Graev, *Comp. Math.*, *35*:299 (1977).
23. A. M. Vershik, I. M. Gelfand, M. I. Graev: *Comp. Math.*, *42*:217 (1982).
24. N. R. Wallach: *Comp. Math.*, *64*:3 (1987).

4

Energy Representations of Path Groups

4.1 INTRODUCTION

In Chapter 3 we showed how to obtain in certain cases irreducibility results for the basic energy representations (Theorem 4). Such results solve the problem of exhibiting candidates for a noncommutative (multiplicative) distribution in the sense of the introduction to Chapter 1. As noticed in Chapter 3, the study of energy representations of $\mathfrak{D}(X, G)$, X a Riemanian manifold and G a compact semisimple Lie group, is far from being complete for the critical dimension $\dim X = 2$ when the roots of the Lie algebra of G have a small length. In fact for this case there are just indications in the direction of reducibility of the energy representation but nothing explicit is known about elements appearing in a possible decomposition of the representation in irreducible components.

The aim of this chapter is to show that, on the other hand, the case $\dim X = 1$ is under good control: reducibility of the energy representation is proven and the irreducible components have been constructed [2]. The main idea, first pointed out in [1], is to realize that in the case $X = \mathbb{R}$ (resp. $X = \mathbb{R}^+$, resp. $X = S^1$) the energy representation can be given exploiting the quasi-invariance of the standard Wiener measure on the space of continuous maps from X to G. This permits not only to show the reducibility of the representation but also to find the irreducible components, (the latter however, for the time being, only for $G = SU(n)$). The treatment of the one-dimensional case suggests that a possible proof of reducibility for the case dim $X = 2$ with roots shorter than a certain critical value, left open by the analysis of Chapter 3, might come from a construction of certain two-dimensional group-valued (generalized) random fields (which could be viewed as a probabilistic construction of a quantized "nonlinear σ-model").

We also discuss in this section the representation of equivariant loops, which gives an interesting connection with the theory of the representations of Kac-Moody groups and algebras, the object of study in the next chapter.

The present chapter is organized as follows: In Section 1, we introduce the basic objects for the new description of the energy representation for $X = \mathbb{R}$, $\mathbb{R}^+$, or S^1: the standard Brownian measure on the set of paths in G, with a possible restriction on the equivariant paths. In Section 2, a natural decomposition associated to a maximal torus in G is obtained. In this section, we introduce a new kind of representations which are nonlocated and of order 1. In Section 3, irreducibility of these representations and connected results on the energy representation are proved.

The material of this chapter is based essentially on the papers [2, 3, 15].

4.2 STANDARD BROWNIAN MEASURE ON PATH SPACES

Let G be a semisimple compact Lie group and $\mathcal{G}$ its Lie algebra, considered as an Euclidian space with its natural scalar product, denoted by (ξ_1, ξ_2), given by the opposite of the Killing form. Let us consider the generalized stochastic process $\xi(t)$ [7] which is the standard white noise process on $\mathcal{G}$, i.e., for any $f \in L_2(\mathbb{R}, \mathcal{G})$, we define

$$\langle \xi, f \rangle = \int (\xi(t), f(t))\, dt \tag{4.1}$$

as a Gaussian random variable with zero mean and variance

$$|f|_2^2 = \int dt\, (f(t), f(t)) \tag{4.2}$$

The standard Wiener process starting at $t = 0$ at 0 is then given by

$$W(t) = \int_0^t \xi(s)\, ds \tag{4.3}$$

The process $W(t)$ is continuous for almost all realizations, hence defines a measure μ_W on $C(\mathbb{R}^+; \mathcal{G})$, the space of continuous functions from the positive real line $\mathbb{R}^+$ into $\mathcal{G}$.

The *standard Brownian motion on G starting at* $h \in G$: $\eta(t, h)$ is then defined as the solution of the stochastic differential equation

$$d\eta \cdot \eta^{-1}(t) = \xi(t), \qquad t \geq 0 \tag{4.4}$$

such that $\eta(0, h) = h$.

Another way of describing the process $\eta(t, h)$ is to consider it as the solution of the system of equations.

$$\begin{aligned} \int_0^t d\eta\,(\tau, h) \cdot \eta^{-1}\,(\tau, h) &= W(t) \\ \eta(0, h) &= h \end{aligned} \tag{4.5}$$

Using the classical theory of stochastic differential equations [12], one sees that $\eta(t, h)$ is continuous in t for almost all realizations of the process. This gives rise to a measure denoted $\mu_{\mathbb{R}^+}(h)$ on $C(\mathbb{R}^+, G)$, which is the image measure of the Wiener measure on $C(\mathbb{R}^+, \mathcal{G})$ by the bimeasurable mapping from $C(\mathbb{R}^+, g)$ onto $C(\mathbb{R}^+, G)$ given by (4.5). $\eta(t, h)$ is a Markov process and moreover the process $t \to \eta(t + s, h)$ converges in probability whenever $s \to \infty$ to a homogeneous process independent of h, denoted $\eta(t)$ and which is called the *standard Brownian motion on G*. $\eta(t)$ satisfies (4.4). The corresponding measure on $C(\mathbb{R}, G)$ is denoted by μ_R and called the standard Brownian measure on $C(\mathbb{R}, G)$.

Let $S^1 = \mathbb{R}/\mathbb{Z}$ and let μ_{S^1} be the probability measure obtained from μ_R by conditioning with respect to the condition

$$\eta(0) = \eta(1) \tag{4.6}$$

on paths $\eta \in C(\mathbb{R}, G)$. Henceforth we shall refer to this measure as the standard Brownian measure on $C(S^1, G)$.

Considering now as a supplement of structure that G is equipped with an order-2 automorphism τ (in interesting cases τ will be an outer automorphism of $SU(n)$, $n \geq 2$: from [9] we know that this object always exists and is essentially unique). We shall say that a function in $C(S^1, G)$ (a *loop in G*) is an *equivariant loop* in G with respect to τ if it satisfies

$$\tau(f(\theta)) = f(\theta + \tfrac{1}{2}) \tag{4.7}$$

Let $C^e(S^1, G)$ be the set of equivariant loops in G. Then one can define the standard Brownian equivariant measure $\mu^e_{S^1}$ on $C(S^1, G)$ (with respect to τ) to be the measure obtained from the standard Brownian measure $\mu_{\mathbb{R}}$ on $\mathbb{R}$ through the conditioning given by (4.7). Since τ is of order 2, $\mu^e_{S^1}$ is actually supported by those continuous functions in $C(S^1, G)$ which satisfy (4.7).

In what follows we will call standard Brownian measure μ any of the above-defined probability measures on the different spaces $C(\mathbb{R}^+, G)$ (in that case taking $h = e$, the identity in G) (resp. $C(\mathbb{R}, G)$) (resp. $C(S^1, G)$) (resp. $C^e(S^1, G)$). In order to get unified notations we will denote by I any of the sets $\mathbb{R}^+$, $\mathbb{R}$, S^1.

After these preliminaries, let us go to the heart of this section; we want to describe, using the preceeding notions, the energy representation of $\mathcal{D}(I, G)$ and, in the case $I = S^1$, its restriction to the set $\mathcal{D}^e(S^1, G)$ of equivariant C^∞-loops from S^1 to G.

The basic observation is that for $\psi \in \mathcal{D}(I, G)$ (resp. $\psi \in \mathcal{D}^e(S^1, G)$) the mapping $\eta \to \eta\psi$ (pointwise right multiplication of paths (resp. of equivariant loops)) leaves the standard Brownian measure μ on $\mathcal{D}(I, G)$ resp. ($\mathcal{D}^e(S^1, G)$) quasi-invariant. μ is also invariant by $\eta \to \eta^{-1}$ (pointwise inversion of a path (resp. of an equivariant loop)). It follows that the left pointwise multiplication by a C^∞-path (resp. a C^∞-equivariant loop) leaves μ quasi-invariant.[a]

It follows that the operations $U^R(\psi)$, $U^L(\psi)$, $\psi \in \mathcal{D}(I, G)$ (resp. $\mathcal{D}^e(S^1, G)$) defined by

$$(U^R(\psi)F)(\eta) = \left(\frac{d\mu(\eta\psi)}{d\mu(\eta)}\right)^{1/2} F(\eta\psi) \tag{4.8}$$

$$(U^L(\psi)F)(\eta) = \left(\frac{d\mu(\psi^{-1}\eta)}{d\mu(\eta)}\right)^{1/2} F(\psi^{-1}\eta) \tag{4.9}$$

on elements F in $L^2(C(I, G), \mu)$ (resp. $L^2(C^e(S^1, G), \mu)$) are unitary operators on $\mathcal{H} = L^2(C(I, G), \mu)$ (resp. $L^2(C^e(S^1, G), \mu)$). Clearly $\psi \to U^L(\psi)$ and $\psi \to U^R(\psi)$ are mutually commuting, unitarily equivalent, representations of $\mathcal{D}(I, G)$ (resp. $\mathcal{D}^e(S^1, G)$) having the constant vector $\mathbb{1}$:

$$\mathbb{1}(\eta) = 1 \quad \forall\, \eta \in C(I, G) \quad (\text{resp. } \forall\, \eta \in C^e(S^1, G)) \tag{4.10}$$

as a cyclic vector. From the definition of μ, it follows that the corre-

[a]Recently the quasi-invariance of μ has been further unvestigated and the smoothness of the module of quasi-invariance has been proved by Malliavin and Malliavin in [10], a very interesting contribution to the more general attempt of constructing an infinite-dimensional differential geometry using quasi-sure analysis.

sponding spherical function is

$$\psi \to \exp(-\tfrac{1}{2}\,|d\psi\,\psi^{-1}|_2^2)$$

where $|\ |_2$ is defined in (4.2). This allows us to identify U^R or U^L with the cyclic component of the vacuum of the energy representation of $\mathcal{D}(I, G)$ (resp. of the restriction of $\mathcal{D}^e(S^1, G)$ of the energy representation of $\mathcal{D}(S^1, G)$).

Let us notice, at this point, that, since we are in the case dim I = 1, we do not know whether the energy representation admits the vacuum as a cyclic vector. The same remark is a fortiori valid in the equivariant case. So, the results of this section concern only the cyclic component of the energy representation and we leave open the problem of deciding whether or not the energy representation (resp. its restriction to equivariant loops in the case of $\mathcal{D}(S^1, G)$) is cyclic. For the sake of simplicity, we shall use the name "energy representation" for "cyclic component of the vacuum of the energy representation (resp. of the restriction to equivariant loops of the energy representation)."

4.3 REDUCTION OF THE ENERGY REPRESENTATION

We shall see in this section that the energy representation is reducible: the vector $\mathbb{1}$ is cyclic for U^L and U^R; consequently $\mathbb{1}$ is cyclic and separating for both U^L and U^R. In order to get an insight into the irreducible components of U^L, a natural idea is to select a maximal von Neumann abelian subalgebra of U^R and to diagonalize it. A natural candidate for this algebra is the set $U^R(\mathcal{D}(I, T))''$ (resp. $U^R(\mathcal{D}^e(S^1, T))''$) where T is a maximal torus in G (resp. a maximal τ-invariant torus in G). As we will see in the next section, this program can be achieved at least in the case $G = SU(n)$ $(n \geq 2)$.

The main result of this section is stated in the following theorem. (For the concept of diagonalizable operator used in the statement, see, e.g., [14].)

Theorem 1. Let G be a compact semisimple Lie group, T a maximal torus in G and let I be either $\mathbb{R}_+$, $\mathbb{R}$, or S^1. Then

$$\mathcal{H} = \int^{\oplus} \mathcal{H}^{\alpha}\, d\mu_T(\alpha) \tag{4.10}$$

where, in the decomposition, $U^R(\mathcal{D}(I, T))''$ is the set of diagonalizable operators, μ_T is the standard Brownian measure on $C(I, T)$ and, for almost all $\alpha \in C(I, T)$,

$$\mathcal{H}^{\alpha} = L^2(C(I, G/T), \mu_1)$$

where $C(I, G/T)$ is the set of paths in the right cosets set G/T and μ_1 is the canonical image of μ by the quotient mapping $\eta \to \mathring{\eta}$ from $C(I, G)$ onto $C(I, G/T)$ (the *standard Brownian measure on* $C(I, G/T)$).

Moreover, in the decomposition (4.10), $U^L(\psi)$ is, for any $\psi \in \mathcal{D}(I, G)$, a decomposable operator:

$$U^L(\psi) = \int^{\oplus} U^{\alpha}(\psi)\, d\mu_1(\alpha) \tag{4.11}$$

where, for almost all α with respect to μ_1, U^{α} is the representation of $\mathcal{D}(I, G)$ acting on $\mathcal{H}^{\alpha}$ in the following way:

$$(U^{\alpha}(\psi)F)(\xi) = \left(\frac{d\mu_1\,(\psi^{-1}\xi)}{d\mu_1\,(\xi)}\right)^{1/2} F(\psi^{-1}\,\xi)$$
$$\exp(-i\langle\alpha^{-1}\, d\alpha,\, \Phi^{-1}\, d\psi \cdot \psi^{-1}\,\Phi\rangle). \tag{4.12}$$

Here $\xi \in C(I, G/T)$, $F \in \mathcal{H}^{\alpha} = L^2(C(I, G/T), \mu_1)$, $\psi \in \mathcal{D}(I, G)$, and Φ is (any) representative of the path ξ.

In the equivariant case, this result has the following counterpart:

Theorem 2. Let G be a compact semisimple Lie group with an order-2 automorphism τ and T a maximal torus in G such that $\tau(T) = T$. Then the results of Theorem 1 are valid with the following changes: all sets of paths have to be replaced by sets of equivariant loops and all standard Brownian measures by the corresponding standard Brownian equivariant ones.

Notice, before going into the details of the proofs of Theorems 1 and 2, that the invariance of T by τ allows to consider the quotient action of τ on the cosets spaces (G/T, $T\backslash G$, $T\backslash G/T$) and the notion of equivariant loops with values in these sets makes sense.

We will write in details the proof of Theorem 1 and we shall give indications about the changes necessary in order to get the similar proof for Theorem 2, i.e., for the equivariant case.

The fundamental lemma is the following.

Lemma 1. Consider G as a principal fiber bundle with the right action of T and the connection on G such that the vertical subspace in the tangent space G_x at $x \in G$ is $x\mathcal{T}$, where $\mathcal{T}$ is the Lie algebra of T imbedded in $\mathcal{G}$ and the horizontal subspace at x is $x\mathcal{T}^{\perp}$ where $\mathcal{T}^{\perp}$ is the orthogonal complement of $\mathcal{T}$ with respect to the Euclidean structure in $\mathcal{G}$ given by the opposite of the Killing form. Then:

(i) For almost all $\eta \in C(I, G)$ with respect to the standard Brownian measure, there exists a canonical decomposition (*the horizontal decomposition of* η) $\eta = \Phi\alpha$ where $\alpha \in C(I, T)$ and Φ has horizontal increments, i.e., such that, for any $\delta \in \mathcal{D}(I, \mathcal{T})$, one has

$$\int_I (\delta(\tau), \Phi^{-1}(\tau)\, d\Phi(\tau)) = 0 \tag{4.13}$$

(ii) The mapping $\eta \to (\mathring{\eta}, \alpha)$ where $\mathring{\eta}$ is the canonical image of η in $C(I, G/T)$, is bimeasurable and transforms the standard Brownian measure on $C(I, T)$ into $\mu_1 \otimes \mu_T$.

(iii) If $\eta = \Phi\alpha$ is the horizontal decomposition of η, then the horizontal decomposition of $\psi^{-1}\eta$ is

$$\psi^{-1}\eta = (\psi^{-1}\, \Phi\beta)(\beta^{-1}\alpha),$$

where β is a path in T satisfying the following differential equation

$$\beta^{-1}\, d\beta = P(\Phi^{-1}\, d\psi\, \psi^{-1}\Phi) \tag{4.14}$$

where P is the orthogonal projection from $\mathcal{G}$ into $\mathcal{T}$.

PROOF. The infinitesimal version of (4.13) is the equation

$$P((\eta\alpha^{-1})^{-1}\, d(\eta\alpha^{-1})) = 0$$

in the distribution sense. Equivalently, since $\alpha \cdot \alpha^{-1}$ acts trivially on $L^2(I, t)$, one has to solve

$$\alpha^{-1}\, d\alpha = P(\eta^{-1}\, d\eta) \tag{4.15}$$

Equation (4.15) has a unique solution α in $C(I, T)$ with $\alpha(0) = e$, since its second member is an element of $L^2(I, \mathcal{T})$. Since $\alpha^{-1} \cdot \alpha$ acts unitarily on $L^2(I, \mathcal{G})$, $\alpha^{-1}\varphi^{-1}\, d\varphi\, \alpha$ and $\alpha^{-1}\, d\alpha$ are pointwise orthogonal in $\mathcal{G}$ and, consequently, the mapping $\eta \to (\varphi, \alpha)$ transforms the standard Brownian measure into the probability measure $\mu_h \otimes \mu_T$, where μ_h is the measure on paths with horizontal increments Φ such that $\Phi^{-1}\, d\Phi$ is white noise distributed in $\mathcal{T}^{\perp}$.

The injectivity of the restriction of the canonical map from $C(I, G)$ onto $C(I, G/T)$ to the paths with horizontal increments comes from the fact that α is unique. This map transforms μ_h in μ_1. It is then possible to look at a function in $C(I, G)$ as a function of two variables, i.e., as an element of $C(I, G/T) \times C(I, T)$ with the distribution given in (ii).

Notice that, in the description (ii), the functions which are invariant by right multiplication by elements in $\mathcal{D}(I, T)$ are the functions which do not depend on the second variable. More generally, right translations by elements of $\mathcal{D}(I, T)$ appear as right translations of the second variable. This remark will be of interest in the proof of Theorem 1.

In order to understand the appearance of the term β in (iii), let us first remark that if Φ has horizontal increments, then $\psi^{-1}\Phi$ ($\psi \in \mathcal{D}(I, G)$) has certainly not the same property. So one has to correct, using a path β with vertical increments, the term $\psi^{-1}\Phi$ in order to get the desired property. The horizontality of $\psi^{-1}\Phi\beta$ can be written, in the distribution sense, as

$$P((\beta^{-1}\Phi^{-1}\psi)\, d(\psi^{-1}\Phi\beta)) = 0$$

This is equivalent to Equation (4.14) as seen using horizontality of $\Phi^{-1} d\Phi$ and the unitarity (resp. the triviality) of $\beta^{-1} \cdot \beta$ on $L^2(I, \mathcal{G})$ (resp. on $L^2(I, \mathcal{T})$).

Let us now go over the proof of Theorem 1. We start with the description of the representation U^L, U^R on the set of functions of two variables introduced in (ii) of Lemma 1. We define the unitary operator

$$(WF)(\xi, \alpha) = F(\Phi\alpha)$$

$(F \in L^2(C(I, G), \mu))$, $\xi \in C(I, G/T)$, $\alpha \in C(I, T)$, $\xi = \mathring{\Phi}$, and where (Φ, α) are elements of the horizontal decomposition of the path $\Phi\alpha$. Let us set

$$V^L(\psi) = WU^L(\psi)W^{-1}, \qquad V^R(\psi) = WU^R(\psi)W^{-1}$$

Then we have

$$(V^L(\psi)F)(\xi, \alpha) = \left(\frac{d\mu_1(\psi^{-1}\xi)}{d\mu_1(\xi)}\right)^{1/2} \left(\frac{d\mu_T(\beta^{-1}\alpha)}{d\mu_T(\alpha)}\right)^{1/2} F(\psi^{-1}\xi, \beta^{-1}\alpha) \tag{4.16}$$

for any $\psi \in \mathcal{D}(I, G)$, $\xi \in C(I, G/T)$, $\alpha \in C(I, T)$, and $F \in L^2(C(I, G/T) \times C(I, T), \mu_1 \times \mu_T)$.

As expected, the restriction of U^R to $C(I, T)$ has a very simple expression:

$$(V^R(\gamma)F)(\xi, \alpha) = \left(\frac{d\mu_T(\alpha\gamma)}{d\mu_T(\alpha)}\right)^{1/2} F(\xi, \alpha\gamma)$$

$\gamma \in C(I, T)$, ξ, α, F as before.

We can now perform the simultaneous diagonalization of operators in $U^R(\mathcal{D}(I, T))$ using Fourier transform. The corresponding spectral measure is just the classical Gaussian distribution with mean 0 and unit covariance. Elements in $U^R(\mathcal{D}(I, T))$ are integrals of characters acting multiplicatively:

$$U^R(\beta) = \int^{\oplus} \exp(i(\alpha^{-1}\, d\alpha,\ \beta^{-1}\, d\beta))\, d\mu_T(\alpha) \tag{4.17}$$

Transporting back the structure by the unitary conjugacy by W, one gets the statements of Theorem 1.

The term $\exp(-i(\alpha^{-1}\, d\alpha,\ \Phi^{-1}\, d\psi\ \psi^{-1}\Phi))$ in Equation (4.12) naturally comes from the evaluation of one of the typical characters appearing in the decomposition (4.17) on $\beta^{-1}\, d\beta$, the vertical connection introduced in (iii) of Lemma 1.

Theorem 2 can be proved along the same lines using a lemma corresponding to Lemma 1 in which all paths groups are replaced by groups of equivariant loops and standard Brownian measures by equivariant ones. Minor modifications have to be made in the equivariant context. The main one is connected to the fact that one has to obtain for an equivariant loop η an equivariant horizontal decomposition $\eta = \Phi\alpha$ with α equivariant. This involves the following trick: the equations

$$\alpha^{-1}\, d\alpha = P(\Phi^{-1}\, d\Phi)$$

or

$$\beta^{-1}\, d\beta = P(\Phi^{-1}\psi^{-1}\, d\psi\ \Phi)$$

have equivariant second members with respect to the evident action of τ on the Lie algebra $\mathcal{G}$. It is not clear that a solution in α or β should be equivariant, but from any solution α_0 or β_0 one can build an equivariant solution α or β by

$$\alpha(t) = \alpha_0(t)\tau(\alpha_0(t + \tfrac{1}{2}))$$

$$\beta(t) = \beta_0(t)\tau(\beta_0(t + \tfrac{1}{2}))$$

Let us conclude this section by noticing that Equation (4.12) introduces an order-1 representation of $\mathcal{D}(I, G)$ which is a good candidate for an order-1 noncommutative multiplicative distribution. The question is now to examine whether irreducibility holds and the next section is devoted to the discussion of this question (in the particular case $G = SU(n)$, $n \geq 2$).

4.4 IRREDUCIBILITY OF ALMOST ALL U^α'S AND PROPERTIES OF THE RIGHT AND LEFT VERSIONS U^R, U^L OF THE ENERGY REPRESENTATION

In order to get information about the representations U^α, which appear in the decomposition of the energy representation discussed in the preceding section, we use the method of partial diagonalization of the representation similarly as before, i.e., a diagonalization of a suitably chosen abelian subalgebra. We use the same notions and notations as before.

As in Section 2, the desired decomposition of algebras of operators have an underlying space counterpart whose description is the contents of the first lemma.

Lemma 2. Consider G as a principal fiber bundle with the left-right action of $T \times T$ given by ($\forall x \in G$, $\alpha, \beta \in T$)

$$(\alpha, \beta) \cdot x = \alpha^{-1} x \beta$$

G is equipped with a connection with $\mathcal{T}x + x\mathcal{T}$ as vertical subspace and its orthogonal space in $x\mathcal{G}$ as horizontal subspace. Then:

(i) For μ-almost all path η in $C(I, G)$, there is a decomposition (*the bihorizontal decomposition of* η)

$$\eta = \alpha^{-1} \Phi \beta \tag{4.18}$$

where α, β are continuous, $\alpha(0) = \alpha(\beta) = e$ (the unit in G) and Φ has *bihorizontal increments*, i.e., for any $\delta \in \mathcal{D}(I, T)$

$$\int_I (\delta, \Phi^{-1}\, d\Phi) = \int_I (\delta, d\Phi\, \Phi^{-1}) = 0 \tag{4.19}$$

(ii) The mapping $\eta \to (\alpha, \mathring{\mathring{\eta}}, \beta)$ is a bimeasurable mapping from $C(I, G)$ onto $C(I, T) \times C(I, T\backslash G/T) \times C(I, T)$, where $\mathring{\mathring{\eta}}$ is the image of η by the canonical mapping from $C(I, G)$ onto the set of paths in the double coset space $C(I, T\backslash G/T)$. This isomorphism transports μ into the measure

$$\int \nu^\Phi \, d\mu_b(\Phi) \tag{4.20}$$

where μ_b is a measure supported by paths with bihorizontal increments whose image by the restriction of the canonical mapping $\eta \to \mathring{\mathring{\eta}}$ from $C(I, T)$ onto $C(I, T\backslash G/T)$ is the standard Brownian measure of $C(I, T\backslash G/T)$, and ν^Φ is the measure on $C(I, T \times T) \cong C(I, T) \oplus C(I, T)$ such that the pairs $(\alpha^{-1}\, d\alpha, \beta^{-1}\, d\beta)$ have a Gaussian distribution with mean zero and covariance matrix

$$\begin{bmatrix} \mathbb{1} & -P \circ \mathrm{Ad}\, \Phi \\ -P \circ \mathrm{Ad}\, \Phi^{-1} & \mathbb{1} \end{bmatrix} \tag{4.21}$$

Φ runs over $C(I, T \backslash G/T)$.

Clearly this parametrization of $C(I, G)$ was chosen in order to be able to treat easily right (resp. left) translations by elements γ (resp. γ^{-1}) in $C(I, T)$. These mappings are described as follows:

$$(\alpha, \overset{\circ\circ}{\eta}, \beta) \to (\alpha, \overset{\circ\circ}{\eta}, \beta\gamma) \qquad (\text{resp. } (\alpha, \overset{\circ\circ}{\eta}, \beta) \to (\alpha\gamma, \overset{\circ\circ}{\eta}, \beta))$$

PROOF OF LEMMA 2. In a distributional sense, Equations (4.19) are

$$P(\Phi^{-1}\, d\Phi) = P(d\Phi\, \Phi^{-1}) = 0$$

or, equivalently,

$$\begin{aligned} \alpha^{-1}\, d\alpha - P(\eta\beta^{-1}\, d\beta\, \eta^{-1}) &= P(d\eta\, \eta^{-1}) \\ \beta^{-1}\, d\beta - P(\eta^{-1}\alpha^{-1}\, d\alpha\, \eta) &= P(\eta^{-1}\, d\eta) \end{aligned} \tag{4.22}$$

These equations can be solved for α and β with $\alpha(0) = \beta(0) = e$ for almost all η, because μ-almost surely $(\eta\mathcal{T}\eta^{-1}) \cap \mathcal{T} = \{0\}$ as a consequence of the semisimplicity of G and of the Brownian character of paths, implying the invertibility for almost all μ of the matrix

$$\begin{bmatrix} \mathbb{1} & -P \circ \mathrm{Ad}\, \eta \\ -P \circ \mathrm{Ad}\, \eta^{-1} & \mathbb{1} \end{bmatrix}$$

This argument also gives the uniqueness of the pair α, β under the stated conditions.

The mapping $\eta \to (\alpha, \Phi, \beta)$ is thus one to one and transforms T-left and right invariant functions into functions which only depend on Φ. Therefore, the conditioning with respect to $\overset{\circ\circ}{\eta}$ is just obtained by specifying a given value of Φ. Finally, the statement about measures (ii) is a consequence of the following equation:

$$\begin{aligned} |\eta^{-1}\, d\eta|_2^2 = {} & |\alpha^{-1}\, d\alpha|_2^2 + |\beta^{-1}\, d\beta|_2^2 \\ & - 2(\Phi^{-1}\alpha^{-1}\, d\alpha\, \Phi, \beta^{-1}\, d\beta) + |\Phi^{-1}\, d\Phi|_2^2 \end{aligned}$$

which follows from (4.18) using bihorizontality of increments of Φ. The term $|\Phi^{-1}\, d\Phi|_2^2$ is responsible for the appearance of μ_b in (4.20) and the other terms correspond to the quadratic form associated with the matrix (4.21), which is clearly positive definite for almost all Φ.

Lemma 2 has a counterpart in the case of equivariant loops with respect to τ. Each path group has to be replaced by the corresponding set of equivariant loops. In particular, one chooses T to be τ-invariant and this allows one to consider equivariant loops in $C(S^1, T\backslash G/T)$ with the quotient action of τ on $T\backslash G/T$. For the proof of the lemma in the equivariant case, one has to apply the same trick as in the proof of Lemma 1 in order to get equivariant solutions for α, β of Equations (4.22) which in that case appear with all their coefficients equivariant in the convenient sense.

An easy corollary of Lemma 2 is

Corollary 1. The restrictions of U^L and U^R to the group of smooth paths in $\mathfrak{D}(I, T)$ are direct integral representations

$$\begin{aligned} U^L(\alpha) &= \int^{\oplus} U^{L,\Phi}(\alpha)\, d\mu_b(\Phi) \\ U^R(\alpha) &= \int^{\oplus} U^{R,\Phi}(\alpha)\, d\mu_b(\Phi) \end{aligned} \tag{4.23}$$

$\forall\, \alpha \in \mathfrak{D}(I, T)$.

The representation of $(\alpha, \beta) \in \mathfrak{D}(I, T \times T) = \mathfrak{D}(I, T) \oplus \mathfrak{D}(I, T)$ given by

$$(\alpha, \beta) \to U^{L,\Phi}(\alpha) U^{L,\Phi}(\beta)$$

is Gaussian, acts on $L^2(C(I, T \times T))$, has the function $\mathbb{1}(\alpha, \beta) = 1$ ($\forall\, \alpha, \beta \in C(I, T)$) as a cyclic vector and its corresponding spherical function is

$$\begin{aligned} (\mathbb{1}, U^{L,\Phi}(\alpha) U^{R,\Phi}(\beta)\mathbb{1}) = \exp[-\tfrac{1}{2}\,(|\alpha^{-1}\, d\alpha|_2^2 + |\beta^{-1}\, d\beta|_2^2 \\ - 2(\Phi^{-1}\alpha^{-1}\, d\alpha\, \Phi, \beta^{-1}\, d\beta))] \quad \square \end{aligned} \tag{4.24}$$

In the case of $G = SU(n)$, we are able to prove that a very simple parametrization of the space of paths with bihorizontal increments can be obtained by just taking its image in $C(I, T\backslash G/T)$. This will allow us to give a more explicit version of the decomposition of Corollary 1. The main result in this context is the following.

Proposition 1. Let T, T' be maximal tori in $G = SU(n)$, $n \geq 2$. Then, there exists a conull set $\mathcal{N}$ in $C(I, G)$ with respect to the standard Brownian measure μ such that the conditions $\psi, \Phi \in \mathcal{N}$ and

$$(\psi^{-1}\alpha^{-1}\, d\alpha\, \psi, \beta^{-1}\, d\beta) = (\Phi^{-1}\alpha^{-1}\, d\alpha\, \Phi, \beta^{-1}\, d\beta) \tag{4.25}$$

valid for any functions α, β with L^2-logarithmic derivatives and values in

T, T' respectively, imply that ψ and Φ define the same double coset in $T\backslash G/T'$.

An algebraic result connected to the infinitesimal version of Proposition 1 is the following Lemma 3. Let us first introduce some notations.

Definition 1. Let $\mathcal{T}_1$ be a Cartan subalgebra in the Lie algebra $su(n)$ of $SU(n)$ and let us express any $u \in SU(n)$ by a matrix $\{u_{ij}|i, j = 1, \ldots, n\}$ in a basis in which $\mathcal{T}_1$ is diagonal. We say that $u \in SU(n)$ satisfies the *genericity condition* $G(\mathcal{T}_1)$ if for any proper subset A of $\{1, \ldots, n\}$, the vectors u_A^k with coordinates $\{u_{ik}|i \in A\}$ are not orthogonal.

We denote by $O(\mathcal{T}_1)$ the set of elements in $SU(n)$ satisfying the property $G(\mathcal{T}_1)$.

We say that a *Cartan subalgebra $\mathcal{T}_2$ is in generic position with respect to $\mathcal{T}_1$* if $\mathcal{T}_2 = u\mathcal{T}u^*$ for some element u in $SU(n)$ satisfying the condition $G(\mathcal{T}_1)$.

The notion of genericity introduced for $\mathcal{T}_1$, $\mathcal{T}_2$ implies $\mathcal{T}_1 \cap \mathcal{T}_2 = \{0\}$, see below, but is stronger in general, as easy counterexamples show, in case $n \geq 4$.

Notice that $O(\mathcal{T}_1)$ is a connected subset having at least codimension 2 in $SU(n)$: in the case $n = 2$, $O(\mathcal{T}_1)$ is nothing but the complement of the torus T_1 generated by $\mathcal{T}_1$ and, for $n > 2$, $O(\mathcal{T}_1)$ is the complement in $SU(n)$ of a finite number of real dimension 2 subspaces in the matrix description used in Definition 1.

Lemma 3. Let $\mathcal{T}_1$, $\mathcal{T}_2$ be Cartan subalgebras in $su(n)$. If $\mathcal{T}_2$ is in generic position with respect to $\mathcal{T}_1$, then $\mathcal{T}_1$, $\mathcal{T}_2$, $[\mathcal{T}_1, \mathcal{T}_2]$ generate linearly $su(n)$.

PROOF. Let us first remark that, if u_{ik} are matrix elements of an element in $O(\mathcal{T}_1)$ in a basis where $\mathcal{T}_1$ is diagonal, then the condition

$$\sum_{k=1}^{n} u_{ik}\bar{u}_{jk}v_k = 0 \tag{4.26}$$

valid for $i \leq q$ and $j > q$ and a sequence v_k of real numbers implies that v_k is independent of k.

Actually, multiplying (4.26) by u_{jl} and summing with respect to j one gets

$$\sum_{j=1}^{n} u_{jl} \sum_{k=1}^{n} u_{ik}\bar{u}_{jk}v_k = 0$$

By the unitarity of u, one obtains

$$\sum_{j=1}^{q} \alpha_{ij}u_{jl} = v_l u_{jl}$$

with $\alpha_{ij} = \Sigma_{k=1}^{n} u_{ik}\bar{u}_{jk}\nu_k$ the elements of a self-adjoint matrix. In other words $(u_{jl})_{1\le j\le q}$ are eigenvectors corresponding to the eigenvalues ν_l. (Notice that $G(\mathcal{T}_1)$ implies that $u_{ij} \neq 0$, $\forall\ i, j$). If ν_l is not independent of l, we get a contradiction to $G(\mathcal{T}_1)$.

In order to prove Lemma 3, observing that $\mathcal{T}_1$ and $\mathcal{T}_2$ are orthogonal to $[\mathcal{T}_1, \mathcal{T}_2]$ with respect to the Killing form, we have only to prove that $\mathcal{T}_1 \cap \mathcal{T}_2 = \{0\}$ and $\dim[\mathcal{T}_1, \mathcal{T}_2] = (n - 1)^2$. Let us write $\mathcal{T}_2 = u\mathcal{T}_1 u^*$ with u satisfying $G(\mathcal{T}_1)$; assuming $\lambda \in \mathcal{T}_1$, $\mu = u\nu u^* \in \mathcal{T}_2$, with λ, ν diagonal matrices with $\Sigma\lambda_k = \Sigma\nu_k = 0$ in order to satisfy $\lambda = \mu$, one gets Equation (4.26) for $q = 1$. This proves that $\mathcal{T}_1 \cap \mathcal{T}_2 = \{0\}$.

In order to prove $\dim[\mathcal{T}_1, \mathcal{T}_2] = (n - 1)^2$, since $\lambda, \mu \to [\lambda, \mu]$ is bilinear on $\mathcal{T}_1 \times \mathcal{T}_2$, it will be sufficient to prove that $\lambda \in \mathcal{T}_1$, $\mu \in \mathcal{T}_2$, and the condition ($\forall\ x \in su(n)$)

$$\mathrm{Tr}([\lambda, \mu]x) = 0 \quad \text{implies} \quad \lambda = 0 \text{ or } \mu = 0.$$

In order to prove this result, first choose for x roots with respect to $\mathcal{T}$, of the form e_{ij}, $i \le q, j > q$. If, for instance $\lambda \neq 0$ one gets condition (4.26). As a consequence μ is constant; hence $\mu = 0$.

We can go now to the proof of Proposition 1. Let u_0 be such that $T' = u_0 T u_0^*$ and let $\mathcal{N}_{TT'}$ be the set of paths ξ such that, for almost all s in I, $\xi(s) \in u_0^*\ O(\mathcal{T})$ where $O(\mathcal{T})$ is as in Definition 1. Since $O(\mathcal{T})$ is a connected open set complementary to a set of codimension ≥ 2, well-known properties of Brownian paths imply that $\mathcal{N}_{TT'}$ has full μ measure in $C(I, T)$ [13, Chapter 1, Proposition 2.5].

The functions on $O(\mathcal{T}_1)$ of the form $g \to (g^{-1}\alpha g, \beta)$ with $\alpha, \beta \in \mathcal{T}$ have for derivatives at g_0 functions of the form $x \to ([g_0^{-1}\alpha g_0, x], \beta)$. On the other hand, the same functions are constant on each double coset, which is a submanifold of codimension $(n - 1)^2$. For paths in $\mathcal{N}_0$, $g(0)$ is in $u_0^*\ O(\mathcal{T})$ and, in this condition, the derivatives generate also a $(n - 1)^2$ $(= n^2 - 1 - 2(n - 1))$-dimensional subspace, and the proposition is proved.

We remark that for $G \neq SU(n)$, a simple counting of dimensions shows that Lemma 3 cannot be true. This is the reason for limiting ourselves to the case $G = SU(n)$, $n \ge 2$. Whether Proposition 1 holds or not for more general semisimple compact Lie groups is a problem which remains open.

Corollary 2. (i) There is a conull set $\mathcal{N}$ in $C(I, T\backslash G/T)$ with respect to the standard Brownian measure μ_b on $C(I, T\backslash G/T)$ such that the functions

$$\rho \to \langle \Phi^{-1}\alpha^{-1}\, d\alpha\, \Phi, \beta^{-1}\, d\beta \rangle \tag{4.27}$$

where Φ is any representative of ρ and α, β are T-valued functions with L^2-logarithmic derivatives, generate the σ-algebra of Borel sets in $\mathcal{N}$.

(ii) There is a conull set $\mathcal{N}'$ in $C(I, T\backslash G)$ with respect to the standard Brownian measure $\mathring{\mu}$ such that the functions

$$\xi \to \langle \Phi^{-1}\psi^{-1}\alpha^{-1}\, d\alpha\, \psi\Phi, \beta^{-1}\, d\beta\rangle \tag{4.28}$$

(where Φ is any representative of ξ and where α, β, ψ are functions with L^2-logarithmic derivatives having values in T, T, and G respectively) generate the σ-algebra of Borel subsets of $\mathcal{N}'$.

(i) is simple, taking $T = T'$ and $\mathcal{N}$ to be the image of the set $\mathcal{N}_{TT}$ of Proposition 1.

(ii) follows from (i) and from the conjugacy of tori in G together with the fact that the tori of G generate G.

The next step is devoted to obtain a "uniform disjointness" result about Gaussian measures in the spirit of property Γ in Chapter 3. We remark that this result is stronger than classical results (e.g., [5],[8]) on disjointness of Gaussian measures. The first lemma was proved originally in [3]. We state it in a form convenient to our object, and for the sake of completeness, we give a short proof.

Lemma 4. For any s in a bounded open interval J let $A(s)$, $B(s)$ be two $k \times k$ matrices which are self-adjoint and positive definite. Let $A_{ij}(s)$, $B_{ij}(s)$ be the matrix elements of $A(s)$ and $B(s)$ and assume that, $\forall\, s \in J$, $A_{11}(s) \leq \lambda < \lambda' \leq B_{11}(s)$.

Let μ_A and μ_B be the Gaussian measures on $\mathcal{D}'(J, \mathbb{R}^k)$ (the vectorial $\mathbb{R}^k$-valued distributions on J) with covariances

$$\int_J (\psi_{(s)}, A_{(s)}\psi_{(s)})\, ds \text{ and } \int_J (\psi_{(s)}, B_{(s)}\psi_{(s)})\, ds \ (\psi \in \mathcal{D}(J, \mathbb{R}^k)).$$

Then μ_A and μ_B are "uniformly disjoint", i.e., there exists a Borel set $\mathcal{U}$ in $\mathcal{S}'(J, R^k)$ depending only on λ, λ' such that $\mu_A(\mathcal{U}) = 0$ and $\mu_B(\mathcal{U}) = 1$.

Let us first remark that if Δ is the ball of center 0 and of radius R in $\mathbb{R}^2$ and μ_a and $\mu_{\lambda\mathbb{1}}$ are the Gaussian measures with covariance matrices $a \in M_2(\mathbb{R})$ and $\lambda\mathbb{1}$ with $a \leq \lambda\mathbb{1}$, then one has

$$\mu_a(\Delta) \geq \mu_{\lambda\mathbb{1}}(\Delta)$$

and, if $a \geq \lambda'\mathbb{1}$,

$$\mu_a(\Delta) \leq \mu_{\lambda'\mathbb{1}}(\Delta)$$

Using an arbitrary basis in $\mathcal{D}'(J, \mathbb{R}^k)$, one can assume that the measures are realized on $\Omega = (\mathbb{R}^k)^{\mathbb{N}}$ and one can describe sets in Ω by prescriptions on the canonical projection $x_{i,j}$, $i = 1, \ldots, k, j \in \mathbb{N}$. One has

$$\mu_A(\{|x_{k,1}|^2 + |x_{l,1}|^2 \leq R^2\}) \geq \mu_{\lambda\mathbb{1}}\{|x_{k,1}|^2 + |x_{l,1}|^2 \leq R^2\}$$
$$= 1 - e^{-R^2/2\lambda}$$
$$\mu_B(\{|x_{k,1}|^2 + |x_{l,1}|^2 \leq R^2\}) \leq \mu_{\lambda'\mathbb{1}}\{|x_{k,j}|^2 + |x_{l,1}|^2 \leq R^2\}$$
$$= 1 - e^{-R^2/2\lambda'}$$

Choosing a sequence of real numbers R_n such that

$$e^{-R_n^2/2\lambda'} = n^{-1}$$

and choosing the sets

$$\mathcal{U}^k = \{|x_{2n,1}|^2 + |x_{2n+1,1}|^2 \leq R_{n^k} \mid \forall n \in N\}$$

one easily verifies that $\mathcal{U} = \cup_{k\geq 1}\mathcal{U}^k$ satisfies the statement of the lemma. □

We are now able to state and prove the "uniform disjointness result" which is the crucial point in the proof of Theorem 3.

Proposition 2. There exists a conull set $\mathcal{N}'$ in $C(I, T\backslash G/T)$ and a basis $\mathcal{B}$ for the Borel sets in $\mathcal{N}'$ such that for $\forall \in \mathcal{B}$ the measures $\int_A \nu^\rho \, d\mu(\rho)$ and $\int_{CA} \nu^\rho \, d\mu(\rho)$, where CA is the complement of A, are mutually disjoint.

PROOF. By Corollary 2, the Borel structure on $C(I, T\backslash G/T)$ can be generated by sets of $\ddot{\Phi}$ given by conditions of the form $f(\Phi(s)) < d$, with $s \in I$, $d \in \mathbb{R}$, and f matrix elements of P Ad $\Phi(s)P$. Let A be such a set. By almost sure continuity of the trajectories, we have that there exists an open set U, a number $\eta > 0$, a pair of indices i, j such that, for $\Phi \in A$, $\Phi' \in CA$,

$$(P \text{ Ad } \Phi(s)P)_{ij} \leq d - \eta < d < (P \text{ Ad } \Phi'(s)P)_{ij}$$

for $s \in U$. Using a unitary conjugacy which makes a partial diagonalization of the submatrix with entries $(i, i)(i, j)(j, i)(j, j)$, one obtains for the covariances of ν_ρ and $\nu_{\rho'}$ with $\rho = \mathring{\Phi}$, $\rho' = \mathring{\Phi}'$ matrices satisfying conditions of Lemma 4. This shows that the measures in the statement of Proposition 2 have disjoint projections onto the set of $(T \times T)$-valued functions with logarithmic derivatives which are distributions with support in O, which gives the result. □

We are now in position to state and prove the main result of this section.

Theorem 3. Assume $G = SU(n)$, $n \geq 2$, T a maximal torus in G. U^R, U^L (resp. U^α) are the representations of $\mathcal{D}(I, T)$ acting on $L^2(C(I, T), \mu)$ (resp. $L^2(C(I, G/T), \mathring{\mu})$) defined as in (4.8), (4.9) (resp. 4.12). Then

(i) $U^R(\mathcal{D}(I, T))''$ (resp. $U^L(\mathcal{D}(I, T))''$) is maximal abelian in $U^R(\mathcal{D}(I, G))''$ (resp. $U^L(\mathcal{D}(I, G))''$).

(ii) U^R and U^L are factor representations generating von Neumann algebras which are the commutant of each other.

(iii) U^α is irreducible for almost all α with respect to the measure μ_T of the Brownian motion on T.

The proof is obtained by using simple consequences of Corollary 2 and Proposition 2. □

Lemma 5. The representation of $\mathcal{D}(I, T \times T)$ in $L^2(C(I, G), \mu)$ given by

$$(\varphi, \psi) \to U^L(\varphi)U^R(\psi) \tag{4.29}$$

has a simple spectrum. The representation space can be identified with the set of L^2-functions on $(\mathcal{T} \times \mathcal{T})$-valued measures on I with respect to the measure ν_0 whose Fourier transform is given by

$$\int e^{i(f,\chi)}e^{i(g,\chi')}\, d\nu_0(\chi, \chi') = e^{-|f|^2/2}e^{-|g|^2/2} \int C(I, T\backslash G/T) e^{-(f,\eta g \eta^{-1})}\, d\mu(\eta)$$

PROOF. Following Corollary 2, one can identify double cosets of η and the bihorizontal part of η as in Lemma 2 and μ_b with $\mathring{\mathring{\mu}}$. The Gaussian character of the representation $(\alpha, \beta) \to U^{L,\Phi}(\alpha)U^{L,\Phi}(\beta)$ (see Corollary 1) and Proposition 2 prove the cyclicity of $\mathbb{1} \in L^2(C(I, G), \mu)$ (see 4.10) with respect to the representation (4.29). This proves the first part of the lemma. The assertion concerning measures easily follows from the cyclicity and the fact that $(f, \eta g \eta^{-1})$ only depends on the class of η in $C(I, T\backslash G/T)$.

Lemma 6. (i) For almost all α in $C(I, T)$ with respect to the standard Brownian measure and any $\gamma \in \mathcal{D}(I, T)$, the operator

$$(\mathcal{W}(\gamma)f)(\xi) = \left(\frac{d\mu(\gamma^{-1}\xi)}{d\mu(\xi)}\right)^{1/2} f(\gamma^{-1}\xi)$$

is in the von Neumann algebra generated by $U^\alpha(\mathcal{D}(I, T))$.

(ii) The von Neumann algebra generated by $U^\alpha(\mathcal{D}(I, G))''$ contains all operators of multiplication by functions of ξ and all operators of left translation by elements of $\mathcal{D}(I, G)$.

PROOF. The decomposable operator $\mathcal{W}(\gamma)$ which is the Hilbert integral of the (constant) field of operator $\alpha \to \mathcal{W}(\gamma)$ commutes with $U^R(\mathcal{D}(I, T))$. Since the function $\xi \to (\alpha^{-1}\, d\alpha, \xi^{-1}\, d\psi\, \psi^{-1}\xi)$ is invariant by left multiplication by $\gamma \in \mathcal{D}(I, T)$, $\mathcal{W}(\gamma)$ also commutes with $U^\alpha(\mathcal{D}(I, T))$ for almost all α. By Lemma 5, $\mathcal{W}(\gamma)$ is in the von Neumann algebra generated by the representation (4.3.12) and (i) follows.

(ii) By (i) for $\psi \in \mathcal{D}(I, G)$

$$U^\alpha(\psi)W(\gamma)U^\alpha(\psi)^{-1} \in U^\alpha(\mathcal{D}(I, G))''$$

and this, with (ii) in Corollary 2 and Definition 2.3 of U^α, implies statement (ii).

Let us now go over the proof of Theorem 3. In order to prove (i), it is sufficient to consider the case of U^R and to use the conjugacy by $\eta \to \eta^{-1}$. If

$$Q \in U^R(\mathcal{D}(I, T))' \cap U^R(\mathcal{D}(I, G))'' \subset U^R(\mathcal{D}(I, T))' \cap U^L(\mathcal{D}(I, G))',$$

then Q is decomposable in the direct integral (4.10)

$$Q = \int Q^\alpha\, d\mu_T(\alpha)$$

and Q^α commutes for almost all α with $U^\alpha(\mathcal{D}(I, G))$. It follows that Q^α is the multiplication by a function which is invariant by left action of $\mathcal{D}(I, G)$. Then Q^α is a constant and Q is diagonalizable in the direct integral (4.10) and $Q \in U^R(\mathcal{D}(I, T))''$. This proves (i).

(iii) follows from (i) and from [14, Theorem 8-32].

(ii) follows from (iii) in the following way. If Z is in the center of $U^R(\mathcal{D}(I, G))$, then, as before, Z is diagonalizable:

$$Z = \int Z^\alpha\, d\mu_\alpha$$

Since Z commutes with $U^R(\mathcal{D}(I, G))$, then Z^α does not depend on α and Z is a scalar. □

Let us remark that the method of proof we used in contrast with the original proof given in [3] uses considerations relative to one fixed torus T in G. Statement of measures are not depending on what case for I we consider. In the equivariant case, the same line of argument applies under the additional condition that the torus we choose is invariant by the given automorphism. Clearly, one has to replace "path groups" by "groups of equivariant loops" in the appropriate sense. This remark allows us to state the following result which we give here without proof, the proof being very similar to the one of Theor. 3.

Theorem 4. Assume $G = SU(n)$, $n \geq 2$, τ is an order-2 automorphism and T a maximal torus in G, invariant by τ. Let U^R, U^L (resp. U^α) be the representations defined in Section 1 of the group $\mathcal{D}^e(S^1, G)$ acting on $L^2(C^e(S^1, T), \mu)$ (resp. $L^2(C^e(S^1, G/T), \mathring{\mu})$). Then

(i) $U^R(\mathcal{D}^e(S^1, T))''$ (resp. $U^L(\mathcal{D}^e(S^1, T))''$) is maximal abelian in $U^R(\mathcal{D}^e(S^1, G))''$ (resp. $U^L(\mathcal{D}^e(S^1, G))''$).

(ii) U^R and U^L are factor representations generating von Neumann algebras which are the commutant of each other.

(iii) U^α is irreducible for almost all α with respect to the measure μ_T of the equivariant Brownian motion on T.

4.5 COMMENTS ON FURTHER DEVELOPMENTS

The construction of a left-invariant Brownian motion giving a quasi-invariant measure on $C(S^1, G)$ for the left action of $\mathcal{D}(S^1, G)$ has allowed to get a left regular unitary representation of this path group. In this context the previous sections are the beginning of a stochastic harmonic analysis on the loop group $\mathcal{D}(S^1, G)$.

By work of Gaveau and Trauber [6], and Daletskii [4] one can construct, when X is a compact manifold, measures on the Sobolev gauge group $W_2^{(n)}(X, G)$, $n > (1/2) \dim (X)$, which are quasi-invariant by the left action of $\mathcal{D}(X, G)$. This gives rise to harmonic analysis on gauge groups, see [11].

REFERENCES

1. S. Albeverio and R. Høegh-Krohn, *Compositio Math.*, *36*:37 (1978).
2. S. Albeverio, R. Høegh-Krohn, and D. Testard, *J. Funct. Anal.*, *57*:49 (1984).
3. S. Albeverio, R. Høegh-Krohn, D. Testard, and A. Vershik, *J. Funct. Anal.*, *51*:115 (1983).
4. Yu. Daletskii, *Uspekhi. Mat. Nauk*, *81*:87 (1983) (transl. *Russ. Math. Surv.* *38:97* (1983)).
5. J. Feldman, *Pacific J. Math.*, *8*:699 (1958).
6. B. Gaveau and Ph. Trauber, *C. R. Acad. Sci. Paris Ser. A291*:575 (1980).
7. I. M. Gelfand and N. Ya Vilenkin, *Generalized Functions*, vol. 4, Academic Press, London (1964) (translation).
8. J. Hajek, *Czech. Math. J.*, *8*:610 (1958).
9. V. Kac, *Funct. Anal. Appl.*, *3*:252 (1969) (transl.).
10. M. P. Malliavin and P. Malliavin, Quasi invariant integration on loop groups, *J. Funct. Anal. 92* (1990), 207–237

11. J. Marion, Outline of harmonic analysis on groups of paths with values in a Sobolev gauge group, in "Stochastic Processes, Physics and Geometry," (S. Albeverio, G. Casati, V. Cattanes, D. Merlini, R. Moresi, eds.), *World Scient.*, Singapore, pp. 575–584 (1990).
12. H. P. McKean, *Stochastic Integrals*, Academic Press, New York (1969).
13. J. Port and C. Stone, *Brownian Motion and Classical Potential Theory*, Academic Press, New York (1978).
14. M. Takesaki, *Theory of Operator Algebras*, Vol. 1, Springer-Verlag, New York (1979).
15. D. Testard, *Stochastic Processes—Mathematics and Physics II* (S. Albeverio, Ph. Blanchard, L. Streit, eds.), Lecture Notes in Math., vol. 1250, Springer, Berlin, pp. 326–341 (1987).

5

The Algebraic Level: Representations of Current Algebras

5.1 INTRODUCTION

The previous chapters were concerned with unitary representations of some infinite-dimensional groups, i.e., the gauge groups. The techniques used were basically functional analytic and probabilistic. Actually, it is also very useful to exploit algebraic techniques, most adapted to the study of the corresponding infinite-dimensional Lie algebras.

In this chapter, we will mainly focus on infinite-dimensional Lie algebras of polynomial mappings from tori into semisimple Lie algebras, and on central extensions of such "current algebras." Such Lie algebras have been studied for a long time separately by quantum field theorists and mathematicians, but it is fair to say that it was only during the last 20 years that mathematicians and physicists discovered their richness and unifying power. Kac and Moody, in the 1960s, introduced and classified a family of (possibly infinite-dimensional) Lie algebras, the so-called Kac-Moody algebras, generalizing Cartan's classification of semisimple Lie algebras. The Kac-Moody algebras can be built (by generators and relations) from a Cartan matrix, and possess a very rich highest weight representation theory. Among them is the class of affine Lie algebras, which can be realized as (untwisted or twisted) loop algebras, i.e., mappings from the circle S^1 into semisimple Lie algebras. A large class of unitary highest weight representations of affine Lie algebras can be classified, and some of them can be explicitly realized, as "vertex operator algebras" [16].

Current algebras of the form Map (M, G) with $\dim(M) > 1$ are no longer Kac Moody algebras (they do not possess a Cartan matrix); nevertheless, it is possible to give a structure theory for some of them (quasisimple Lie algebras), and to develop a unitary highest weight representation theory for them.

Unfortunately, it must be stressed that the connection between the representation theory of the current groups and that of the corresponding Lie algebras is far from being well understood. For instance, at the present time there does not exist any general representation theory for current algebras which would include the differential of the energy representation studied in Chapters 3 and 4. However, some results are available, for example the connection between the continuous tensor product representations of $\mathfrak{D}(M, SU(n, 1))$ (see Chapter 1) and the exceptional representations of $su(n, 1)^{(1)}$ has recently been pointed out [8, 29].

This chapter is organized as follows: we first give a presentation of Kac-Moody theory, referring to Kac's book [35] for a much more detailed and complete presentation, and then describe some particular features and applications of the associated representation theory: the action of Diff(S^1) (Sugawara and G.K.O. constructions) and the vertex operator realizations.

Finally, we describe the quasisimple theory, together with the associated highest weight representation theory.

5.2 AFFINE KAC-MOODY ALGEBRAS

5.2.1 Structure Theory

Kac-Moody Algebras: (Complex)

We will briefly describe here the construction of Kac-Moody algebras, developed in the first chapters of [35], to fix the notations and conventions. The basic tools in this section will be the following:

Definition 1. (i) A *generalized Cartan matrix* (G.C.M.) is a $n \times n$ matrix $A = \{a_{ij}\}_{i,j=0,n-1}$ such that

$a_{ii} = 2, i = 1, \ldots, n.$
$a_{ij} \in \mathbf{Z}$ and $a_{ij} \leq 0$ if $j \neq i$, $i, j = 1, \ldots, n.$
$a_{ij} = 0 \Rightarrow a_{ji} = 0, i, j = 0, \ldots, n - 1.$

(ii) A G.C.M. A is *indecomposable* if there is no partition of $\{1, \ldots, n\}$ into two subsets such that $a_{ij} = 0$ whenever i belongs to the first subset and j belongs to the second subset.

(iii) A G.C.M. A is said to be *symmetrizable* if there exists a diagonal nonsingular matrix D and a symmetric matrix S such that $A = DS$.

Let A be a G.C.M. of dimension n, assumed for simplicity to be indecomposable (one easily sees from the definition of Kac-Moody algebras to be given below that decomposable G.C.M. lead to direct sum of Kac-Moody algebras). The starting point of the construction of Kac-Moody algebras is a realization of $A = \{a_{ij}\}$. Let $(\mathfrak{h}, \mathcal{B}, \check{\mathcal{B}})$ be a triple where $\mathfrak{h}$ is some complex vector space, $\mathcal{B} = \{\alpha_p, \ldots, \alpha_{n-1}\} \subset \mathfrak{h}'$, and $\check{\mathcal{B}} = \{h_0, \ldots, h_{n-1}\} \subset \mathfrak{h}$ are finite subsets such that

$\alpha_0, \ldots, \alpha_{n-1}$ (resp. $h_0, \ldots, h_{n-1}$) are linearly independent.
$\alpha_i(h_j) = a_{ij}$, $i, j = 0, \ldots, n-1$.
n-rank$(A) = \dim(\mathfrak{h}) - n$.

Define the Lie algebra $\tilde{\mathcal{G}}(A)$ to be the complex Lie algebra generated by $\mathfrak{h}$ and the elements $\{e_i, f_i\}_{i=0,\ldots,n-1}$, with relations

$$
\begin{aligned}
&[e_i, f_j] = \delta_{ij} h_i, && i, j = 0, \ldots, n-1\\
&[h_i, e_j] = a_{ji} e_j, && i, j = 0, \ldots, n-1\\
&[h_i, f_j] = -a_{ji} f_j, && i, j = 0, \ldots, n-1\\
&[h, h'] = 0, && h, h' \in \mathfrak{h}
\end{aligned}
$$

$[\cdot, \cdot]$ being the Lie algebra multiplication, i.e., the Lie bracket.

Using a construction due to Chevalley [7] (see [35, Theorem 1.2] for the detailed proof), one can see that as a vector space, $\tilde{\mathcal{G}}(A)$ splits into the direct sum

$$\tilde{\mathcal{G}}(A) = \tilde{\mathfrak{n}}_- \oplus \mathfrak{h} \oplus \tilde{\mathfrak{n}}_+,$$

where $\tilde{\mathfrak{n}}_-$ (resp. $\tilde{\mathfrak{n}}_+$) is the $\tilde{\mathcal{G}}(A)$-ideal freely generated by the f_i's (resp. the e_i's). Moreover, $\tilde{\mathcal{G}}(A)$ has a unique maximal ideal $\mathfrak{r}$ intersecting $\mathfrak{h}$ trivially.

Definition 2. The quotient $\mathcal{G}(A) = \tilde{\mathcal{G}}(A)/\mathfrak{r}$ is the *Kac-Moody algebra* associated with the matrix A. A is called the *Cartan matrix* of $\mathcal{G}(A)$, and n is the rank of $\mathcal{G}(A)$. $\mathfrak{h}$ is the Cartan subalgebra of $\mathcal{G}(A)$, and the e_i, f_i, h_i, $i = 0, \ldots, n-1$, are the *Chevalley generators* of $\mathcal{G}(A)$. $\mathcal{B}$ (resp. $\check{\mathcal{B}}$) is a *root* (resp. *coroot*) *basis* of $\mathcal{G}(A)$ and α_i (resp. h_i) are the *simple roots* (resp. *coroots*) of $\mathcal{G}(A)$.

The following lemma is an easy consequence of Chevalley's construction.

Lemma 1. The Lie algebra generated by e_i, f_i, $i = 0, \ldots, n-1$, $\mathfrak{h}$, and the relations

$$\begin{aligned} [e_i, f_j] &= \delta_{ij} h_i, && i,j = 0, \ldots, n-1 \\ [h_i, e_j] &= a_{ji} e_j, \quad [h_i, f_j] = -a_{ji} f_j, && i,j = 0, \ldots, n-1 \\ [h, h'] &= 0, && h, h' \in \mathfrak{h} \\ [\mathrm{Ad}(e_i)]^{1-a_{ij}} \cdot e_j &= [\mathrm{Ad}(f_i)]^{1-a_{ij}} \cdot f_j = 0, && i,j = 0, \ldots, n-1 \end{aligned}$$

is isomorphic to the derived Lie algebra $\mathcal{G}'(A)$ of $\mathcal{G}(A)$. Moreover, one has

$$\mathcal{G}(A) = \mathcal{G}'(A) + \mathfrak{h}.$$

$\mathcal{G}(A)$ has the following *rootspace decomposition*

$$\mathcal{G}(A) = \sum_{\alpha \in \mathfrak{h}'}^{\oplus} \mathcal{G}_\alpha,$$

$$\mathcal{G}_\alpha = \{g \in \mathcal{G} \text{ s.t. } [h, g] = \alpha(h) \cdot g, \ \forall\, h \in \mathfrak{h}\}.$$

We denote by $\mathcal{R}$ the spectrum of $\mathrm{Ad}(\mathfrak{h})$, called the *root system*, $\mathcal{R}$ splits into the disjoint union $\mathcal{R} = \mathcal{R}_- \cup \mathcal{R}_+$ where

$$\mathcal{R}_+ = \{\alpha \in \mathcal{R} \text{ s.t. } \alpha = \sum_{i=1}^{n} k_i \alpha_i, \ k_i \in \mathbb{N}\} = -\mathcal{R}_-$$

$\mathcal{R}_+$ (resp. $\mathcal{R}_-$) is the set of positive (resp. negative) roots.

Moreover, $\mathcal{G}(A)$ admits the triangular decomposition, inherited from that of $\tilde{\mathcal{G}}(A)$,

$$\mathcal{G}(A) = \mathfrak{n}_- \oplus \mathfrak{h} \oplus \mathfrak{n}_+$$

where

$$\mathfrak{n}_\pm = \sum_{\pm\alpha \in \mathcal{R}_+}^{\oplus} \mathcal{G}_\alpha$$

is the $\mathcal{G}(A)$-subalgebra generated by the e_i's (resp. the f_i's). Note that, as usual,

$$\begin{aligned} [\mathcal{G}_\alpha, \mathcal{G}_\beta] &\subset \mathcal{G}_{\alpha+\beta} && \text{if } \alpha + \beta \in \mathcal{R} \\ &= \{0\} && \text{if } \alpha + \beta \notin \mathcal{R} \end{aligned}$$

The following important result is due to Kac [32].

Lemma 2. Both $\mathcal{G}(A)$ and $\mathcal{G}'(A)$ possess an (n-rank (A))-dimensional center, which is equal to

$$\mathcal{C} = \{h \in \mathfrak{h} \text{ s.t. } \alpha_i(h) = 0, i = 0, \ldots, n-1\}$$

From now on, assume that the G.C.M. A is symmetrizable: $A = DS$, $D = \text{Diag}(d_1, \ldots, d_n)$. Let

$$\mathfrak{h}_1 = \sum_{i=0}^{n-1}{}^{\oplus} h_i \subset \mathfrak{h}$$

let $\mathfrak{h}_2$ be a complementary subspace of $\mathfrak{h}_1$ in $\mathfrak{h}$ and define the **C**-valued bilinear form $(\cdot, \cdot)$ on $\mathfrak{h}$ by

$$(h_i, h) = d_i\alpha_i(h) \qquad \forall\, h \in \mathfrak{h}, i = 0, \ldots, n-1$$

$$(h_2', h_2') = 0 \qquad \forall\, h_2', h_2' \in \mathfrak{h}_2$$

One then has

Proposition 1. (i) $(\cdot, \cdot)$ is symmetric and nondegenerate on $\mathfrak{h}$.

(ii) $(\cdot, \cdot)$ extends in a unique way to a nondegenerate invariant symmetric bilinear form on $\mathcal{G}(A)$, called the *standard invariant form*, or *Killing form*.

(iii) The G.C.M. A is connected with the roots by the relation $A = \{2(\alpha_i, \alpha_j)/(\alpha_i, \alpha_i)\}_{i,j=0,\ldots,n-1}$

The invariance and the nondegeneracy properties of the Killing form have the following immediate consequences.

Corollary. (i) $(\mathcal{G}_\alpha, \mathcal{G}_\beta) = 0$ if $\alpha + \beta \neq 0$.

(ii) The Killing form is nondegenerate on h and on $\mathcal{G}_\alpha \oplus \mathcal{G}_{-\alpha}$, $\alpha \in \mathcal{R}$.

The last result to be stated in this subsection is a classification theorem of the G.C.M. [31, 47, 50].

Theorem 1. Let A be any indecomposable G.C.M.; then it falls into one and only one of the three following classes:

(i) All the principal minors of A are positive; A is of finite type.

(ii) $\text{Det}(A) = 0$, and all the proper principal minors are positive; A is of affine type.

(iii) A is of indefinite type.

REMARK. In the affine case, as well as in the finite case, the G.C.M. is always symmetrizable, which allows the construction of the associated Killing form.

Affine Kac-Moody Algebras: (Complex)

The classification of the finite indecomposable G.C.M., and of the associated Lie algebras, the simple Lie algebras, is a classical work going back to Cartan [6]. Kac [31] and Moody [50] have given an analogous result in the affine case.

Theorem 2. Let A be any indecomposable G.C.M. of affine type.

(i) The Dynkin diagram of $\mathcal{G}(A)$ is one of those listed in Table 1 (recall that the Dynkin diagram associates a vertex to each simple root α_i, $i = 0, \ldots, n-1$, and that the ith and jth vertices are connected by an $a_{ij}\, a_{ji}$-fold arrow pointing toward the jth (resp. the ith) vertex if $a_{ij} < a_{ji}$ (resp. $a_{ji} < a_{ij}$)).

(ii) The labels a_i appearing in Table 1 are the coordinates of the unique vector $\delta = (a_0, \ldots, a_{n-1})$ such that $A \cdot \delta = 0$ and the a_i, $i =$

Table 1

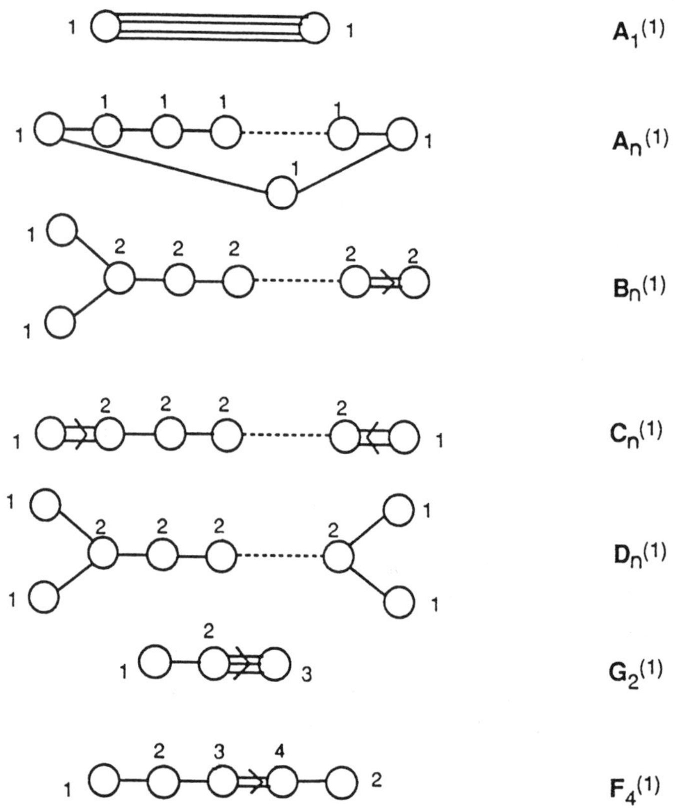

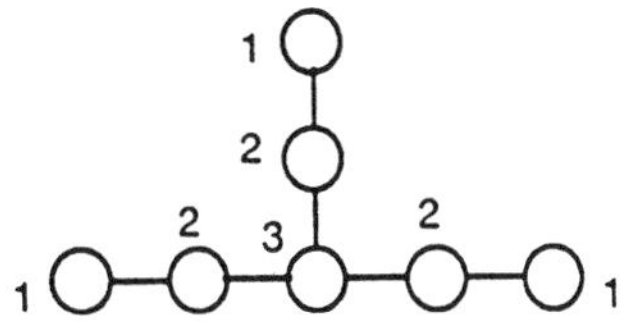

$E_6^{(1)}$

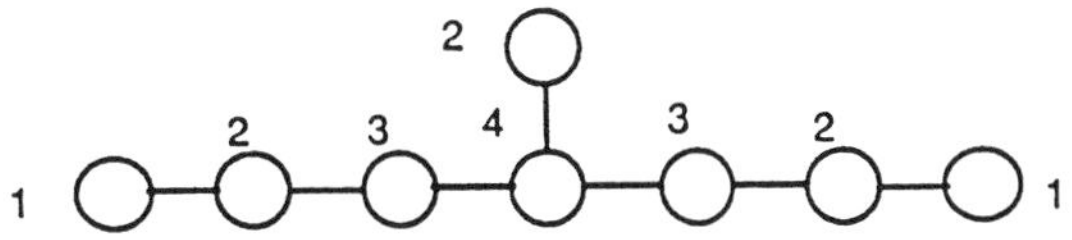

$E_7^{(1)}$

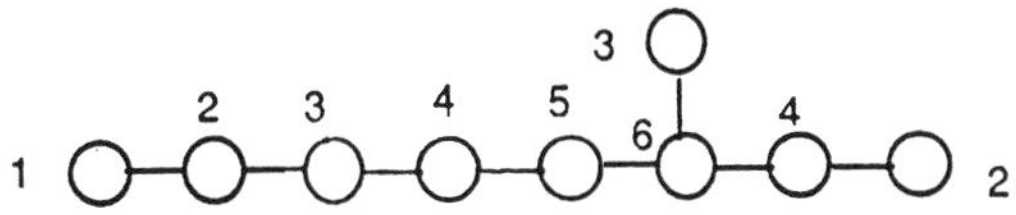

$E_8^{(1)}$

$A_2^{(2)}$

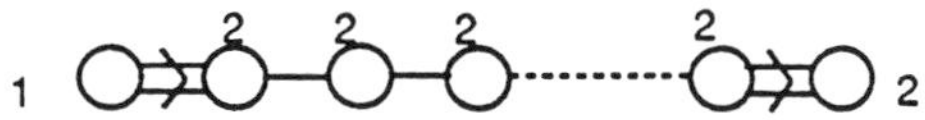

$A_{2l}^{(2)}$

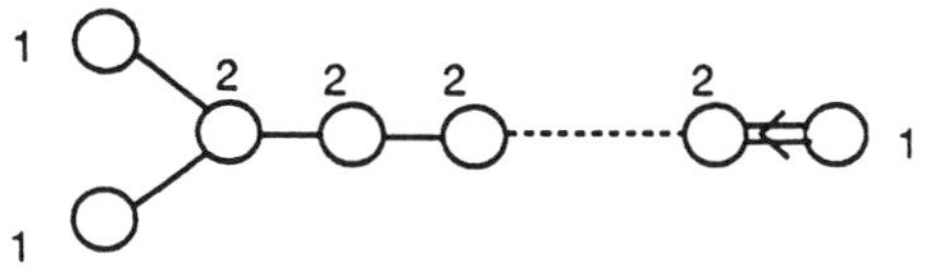

$A_{2l-1}^{(2)}$

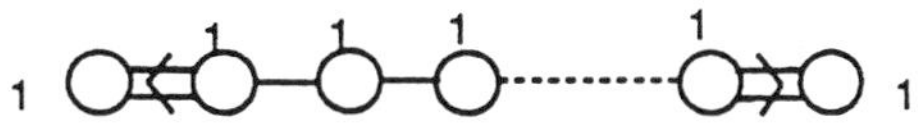

$D_{l+1}^{(2)}$

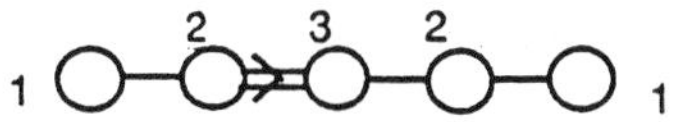

$E_6^{(2)}$

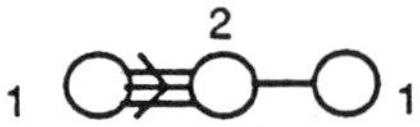

$D_4^{(3)}$

$0, \ldots, n-1$ are relatively prime integer numbers. δ is called the *smallest positive isotropic root* of $\mathcal{G}(A)$.

Let $\mathcal{G}(A)$ be an affine Kac-Moody algebra $\mathbf{X}_N^{(k)}$ of rank $n = l + 1$, and denote by a_i (resp. $\tilde{a}_i$) the numerical labels of its Dynkin diagram (resp. the Dynkin diagram of $\mathcal{G}({}^tA)$, which is obtained by reversing the arrows in the Dynkin diagram of $\mathcal{G}(A)$). They allow us to introduce the *Coxeter* and *dual Coxeter numbers*

$$h = \sum_{i=0}^{l} a_i, \qquad \tilde{h} = \sum_{i=0}^{l} \tilde{a}_i$$

and the *canonical central element* (recall that the center of $\mathcal{G}(A)$ is one-dimensional)

$$c = \sum_{i=0}^{l} a_i h_i$$

By Proposition 1, one can build the (nondegenerate symmetric invariant) Killing form on $\mathcal{G}(A)$ whose restriction to $\mathfrak{h}$ can be described as follows: if $d \in \mathfrak{h}$ is defined by

$$\alpha_0(d) = 1, \qquad \alpha_i(d) = 0, \quad i = 1, \ldots, l$$

then $\mathfrak{h}$ is clearly spanned by $h_0, \ldots, h_l$ and d (or $h_1, \ldots, h_l$, c, and d), and the Killing form is characterized on $\mathfrak{h}$ by

$$(h_i, h_j) = \frac{a_i}{\tilde{a}_i} a_{ji} = \frac{a_j}{\tilde{a}_j} a_{ij}, \qquad i, j = 0, \ldots, l$$

$$(h_i, d) = (d, d) = (c, c) = 0, \qquad i = 1, \ldots, l$$

$$(h_0, d) = (c, d) = 1$$

(note that $a_0 = 1$ for any affine G.C.M.). Denote by $\mathfrak{h}_{\mathfrak{l}}$ the subspace of $\mathfrak{h}$ spanned by $h_1, \ldots, h_{n-1}$ (*gradient Cartan Subalgebra*). Then

$$\mathfrak{h} = \mathfrak{h}_{\mathfrak{l}} \oplus \mathbf{C}c \oplus \mathbf{C}d$$

To carry the Killing form onto $\mathfrak{h}'$ by duality, one first needs to define $\Lambda_0 \in \mathfrak{h}'$ by

$$\Lambda_0(h_i) = 0 \qquad \forall\, i = 1, \ldots, l$$

$$\Lambda_0(h_0) = \Lambda_0(c) = 1$$

The induced symmetric bilinear form, which we call *Killing form* for simplicity, is then characterized by

$$(\alpha_i, \alpha_j) = \frac{\tilde{a}_i}{a_i} a_{ji} = \frac{\tilde{a}_j}{a_j} a_{ij}, \qquad i, j = 0, \ldots, l$$

$$(\alpha_i, \Lambda_0) = (\Lambda_0, \Lambda_0) = (\delta, \delta) = 0, \qquad i = 1, \ldots, l$$

$$(\alpha_0, \Lambda_0) = (\delta, \Lambda_0) = 1$$

For convenience, introduce the *Weyl element* $\rho \in \mathfrak{h}'$, defined by

$$\rho(h_i) = 1 \qquad (i = 0, \ldots, l)$$

$$\rho(d) = 0$$

Note that $\rho(c) = h$. With the help of this nondegenerate bilinear form, one can introduce the notion of real (or nonisotropic) and imaginary (or isotropic or null) roots, and the Weyl reflections.

Definition 3. (i) Let $\alpha \in \mathcal{R}$; then α is an *isotropic* (or *imaginary*) *root* if $(\alpha, \alpha) = 0$; nonisotropic roots are also called *real roots*.

(ii) Let α be any real root in $\mathcal{R}$; then the *Weyl reflection* associated to α (i.e., the reflection with respect to the hyperplane perpendicular to α in $\mathfrak{h}'$) is defined by

$$w_\alpha \cdot \lambda = \lambda - 2\frac{(\alpha, \lambda)}{(\alpha, \lambda)}\alpha, \qquad \lambda \in \mathfrak{h}'.$$

Denote by W the group generated by w_{α_i}, $i = 0, \ldots, n - 1$ (*Weyl group*). The real and imaginary roots have the following crucial properties.

Lemma 3. (i) Let $\alpha \in \mathcal{R}$; then α is real if and only if there exists $w \in W$ and $\alpha_i \in \mathcal{R}$ such that $\alpha = w \cdot \alpha_i$.

(ii) Let ξ be some isotropic root; then $(\xi, \alpha) = 0$ for any $\alpha \in \mathcal{R}$.

Now, like in usual Cartan theory, the affine Kac-Moody algebra can be completely described by the following theorem.

Theorem 3. (i) Let α be any real root. Then

(i) $\dim(\mathcal{G}_\alpha) = 1$.
(ii) $2(\alpha, \beta)/(\alpha, \alpha) \in \mathbf{Z}$ for any $\beta \in \mathcal{R}$.
(iii) $k\alpha \in \mathcal{R}$ if and only if $k = 0$ or ± 1.
(iv) For any $\beta \in \mathcal{R}$: $\beta + n\alpha \in \mathcal{R}$ if and only if $n \in \mathbf{Z}$ and $-n_- \leq n \leq n_+$, where n_- and n_+ are nonnegative integer numbers such that $n_- - n_+ = 2(\alpha, \beta)/(\alpha, \alpha)$.

(ii) Let ξ be any isotropic root in $\mathcal{R}$; then it is of the form $\xi = n\delta$, $n \in \mathbf{Z}$.

REMARKS. 1. If α is a real root, then $\mathcal{G}_{-\alpha} \oplus [\mathcal{G}_\alpha, \mathcal{G}_{-\alpha}] \oplus \mathcal{G}_\alpha$ is isomorphic to $sl(2, \mathbf{C})$; the proof of (i) is standard from usual $sl(2, \mathbf{C})$ representation theory.

2. An easy proof of (ii) can be found in [47]. Let $\mathcal{R}$ be an affine root system of the type $\mathbf{X}_N^{(k)}$. Thanks to Theorem 3, the root system $\mathcal{R}$ can be completely reconstructed from the basis $\mathcal{R}$ of simple roots and the G.C.M. coefficients. One can also give the following simple description of $\mathcal{R}$. First of all, define $\mathcal{R}_{\mathfrak{f}}$ to be the (finite) root system generated by $a_1, \ldots, \alpha_l$ if $\mathcal{R} \neq \mathbf{A}_{2l}^{(2)}$, and by $\alpha_1, \ldots, \alpha_l, 2\alpha_l$ if $\mathcal{R} = \mathbf{A}_{2l}^{(2)}$, $\mathcal{R}_{\mathfrak{f}}$ is the gradient root system.

From usual finite root system theory [5], one knows that if $\mathcal{R} \neq \mathbf{A}_{2l}^{(2)}$ (resp. $\mathcal{R} = \mathbf{A}_{2l}^{(2)}$), one can generically separate the roots into two categories, the long roots α_L and the short roots α_s, such that $(\alpha_L, \alpha_L) = k(\alpha_s, \alpha_s)$ (resp. three categories: the superlong roots $\alpha_{SL} = 2\alpha_s$, the long roots α_L and the short roots α_s, with $(\alpha_{SL}, \alpha_{SL}) = 2(\alpha_L, \alpha_L) = 4(\alpha_s, \alpha_s)$). By Lemma 3(ii), the same property holds for the set $\mathcal{R}_r$ of real roots of $\mathcal{R}$.

REMARK. If $k = 1$, then there is only one category of roots, which have all the same length. $\mathcal{R}_{\mathfrak{f}}$ can also be viewed as the quotient set of $\mathcal{R}$ for the equivalence relation

$$\alpha \sim \beta \quad \text{if } (\alpha, \gamma) = (\beta, \gamma) \quad \forall\, \gamma \in \mathcal{R}, \qquad \alpha, \beta \in \mathcal{R}$$

Proposition 2. The root system $\mathcal{R} = \mathbf{X}_N^{(k)}$ can be described as follows:

Short roots

$$\mathcal{R}_s = \{\alpha_s + n\delta,\ n \in \mathbf{Z},\ \alpha_s \text{ short root of } \mathcal{R}_{\mathfrak{f}}\}.$$

Long roots

$$\mathcal{R}_L = \{\alpha_L + nk\delta,\ n \in \mathbf{Z},\ \alpha_L \text{ long root of } \mathcal{R}_{\mathfrak{f}}\} \qquad \text{if } \mathcal{R} \neq \mathbf{A}_{2l}^{(2)},$$

$$\mathcal{R}_L = \{\alpha_L + n\delta,\ n \in \mathbf{Z},\ \alpha_L \text{ long root of } \mathcal{R}_{\mathfrak{f}}\} \qquad \text{if } \mathcal{R} \neq \mathbf{A}_{2l}^{(2)}.$$

Superlong roots: if $\mathcal{R} = \mathbf{A}_{2l}^{(2)}$

$$\mathcal{R}_{sL} = \{2\alpha_s + (2n+1)\delta,\ n \in \mathbf{Z},\ \alpha_s \text{ short root of } \mathcal{R}_{\mathfrak{f}}\}.$$

Isotropic roots: $\mathcal{R}_i = \{n\delta,\ n \in \mathbf{Z}\}$.

Real Affine Algebras

To define the real affine algebras, one needs to introduce the notion of conjugation on a Lie algebra. If $\mathcal{G}$ is any complex Lie algebra, a *conjugation* of $\mathcal{G}$ is an involutive antilinear antiautomorphism ω of $\mathcal{G}$, i.e. a map ω: $\mathcal{G} \to \mathcal{G}$ such that

$$\omega(\lambda x + \mu y) = \lambda^* \omega \cdot x + \mu^* \omega \cdot y; \quad x, y \in \mathcal{G},\ \lambda, \mu \in \mathbf{C}.$$
$$\omega[x, y] = [\omega \cdot y, \omega \cdot x].$$
$$\omega^2 = 1.$$

Then, if ω is a conjugation of the complex Lie algebra $\mathcal{G}$, the real form $\mathcal{G}\omega$ of $\mathcal{G}$ associated with ω is defined by

$$\mathcal{G}\omega = \{x \in \mathcal{G} \text{ s.t. } \omega \cdot x = -x\}$$

The conjugation ω is said to be *consistent* if $\omega \cdot \mathcal{G}_\alpha = \mathcal{G}_{-\alpha}$ for any root α of $\mathcal{G}$ ($\mathcal{G}_\alpha$ standing as usual for the rootspace attached to the root α).

In the case of affine Kac-Moody algebras, it is fairly easy to see that a conjugation ω is completely determined by its action on the Chevalley generators $\{e_i, f_i\}_{i=0,\ldots,l}$. Moreover, any consistent conjugation of an affine Kac-Moody algebra can be realized as follows [29]

$$\omega \cdot e_i = \pm f_i, \quad \omega \cdot f_i = \pm e_i, \qquad i = 0, \ldots, l$$

As an example, the Cartan conjugation ω_c is defined by

$$\omega_c \cdot e_i = f_i, \quad \omega_c \cdot f_i = e_i, \qquad i = 0, \ldots, l$$

and defines the compact real form $\mathcal{G}_c$ of $\mathcal{G}$.

REMARK. ω extends uniquely to a conjugation on the universal enveloping algebra $\mathcal{U}(\mathcal{G})$ of $\mathcal{G}$ [35].

5.2.2 Loop Algebra (or Current Algebra) Realizations

In addition to the abstract construction of the previous section, all the affine Kac-Moody algebras can be realized as (untwisted or twisted) loop algebras (particular cases of the current algebras of quantum gauge theories, where the space is the circle S^1). First of all, consider the simplest cases, i.e., the untwisted cases.

The Case k = 1

Consider a simple Lie algebra $\mathcal{G}_0$, of the type $\mathbf{X}_N$ (see Table 2), and denote by $\mathbf{C}[t, t^{-1}]$ the algebra of Laurent polynomials in the indeterminate t. The tensor product $\mathcal{G}_1 = \mathbf{C}[t, t^{-1}] \otimes \mathcal{G}_0$ can be endowed with a Lie algebra structure, with the pointwise bracket:

$$\forall x_1, x_2 \in \mathcal{G}_0, \quad \forall P_1, P_2 \in \mathbf{C}[t, t^{-1}] \qquad [P_1 \otimes x_1, P_2 \otimes x_2] = P_1 P_2 \otimes [x_1, x_2]_0$$

Such a Lie algebra is not a Kac-Moody algebra (for example, it has an infinite-dimensional Cartan subalgebra $\mathfrak{h}_1 = \mathbf{C}[t, t^{-1}] \otimes \mathfrak{h}_0$, $\mathfrak{h}_0$ being a Cartan subalgebra of $\mathcal{G}_0$); but it can be transformed into an affine Lie algebra $\mathcal{G} = \mathcal{G}_1 \oplus \mathbf{C}c \oplus \mathbf{C}d$ by considering the following extension:

$$[P_1 \otimes x_1 \oplus \lambda_1 c \oplus \mu_1\, d,\ P_2 \otimes x_2 \oplus \lambda_2 c \oplus \mu_2\, d]$$
$$= P_1 P_2 \otimes [x_1, x_2]_0 + \mu_1 t \frac{dP_2}{dt} \otimes x_2 - \mu_2\, t \frac{dP_1}{dt} \otimes x_1$$
$$+ \Psi(P_1 \otimes x_1, P_2 \otimes x_2)$$

with

$$\Psi(P_1 \otimes x_1, P_2 \otimes x_2) = \frac{1}{2\pi i} (x_1, x_2)_0 \oint_0 \frac{dP_1(t)}{dt} P_2(t)\, dt$$

$(. \, . \, .)_0$ being a Killing form in $\mathcal{G}_0$. One easily checks that:

Lemma 4. $\dim H^2(\mathcal{G}_1, \mathbf{C}) = 1$, and $\Psi \in H^2(\mathcal{G}_1, \mathbf{C})$. In other words, Ψ defines the universal central extension of $\mathcal{G}_1$.

Recall that the complex-valued bilinear form Ψ on $\mathcal{G}_1$ belongs to $H^2(\mathcal{G}_1, \mathbf{C})$ if it satisfies

$$\Psi(x, y) = -\Psi(y, x)$$
$$\Psi(x, [y, z]) + \Psi(z, [x, y]) + \Psi(y, [z, x]) = 0$$

for any $x, y, z \in \mathcal{G}_1$.

The study of the root system of $\mathcal{G}$ will allow the identification of $\mathcal{G}$ with an affine Lie algebra. We first need to state some conventions and notations.

Denote by $\mathcal{B}_0 = \{\beta_1, \ldots, \beta_N\}$ a root basis of $\mathcal{G}_0$, and $\check{\mathcal{B}}_0 = \{h_1, \ldots, h_N\}$ the corresponding coroot basis. Denote by θ the highest root of $\mathcal{G}_0$ (the highest roots of simple Lie algebras can be found in Table 2)

Table 2

A_l	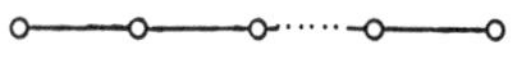	$\theta = \sum_{i=1}^{l} \beta_i$
B_l	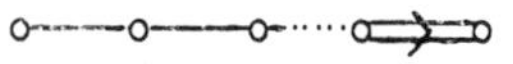	$\theta = \beta_1 + 2 \sum_{i=2}^{l} \beta_i$
C_l	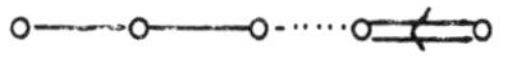	$\theta = \beta_1 + 2 \sum_{i=2}^{l-1} \beta_i + \beta_l$
D_l	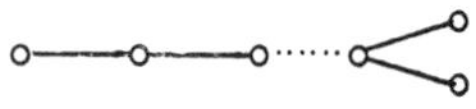	$\theta = 2 \sum_{i=1}^{l-2} \beta_i + \beta_{l-1} + \beta_l$
E_6	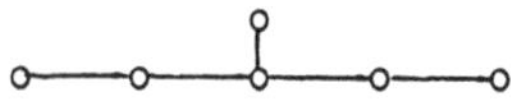	$\theta = \beta_1 + 2\beta_2 + 3\beta_3 + 2\beta_4 + \beta_5 + 2\beta_6$
E_7	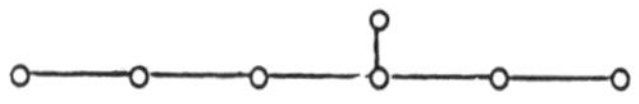	$\theta = \beta_1 + 2\beta_2 + 3\beta_3 + 4\beta_4 + 3\beta_5 + 2\beta_6 + 2\beta_7$

E_8 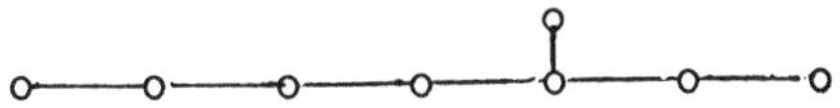 $\theta = 2\beta_1 + 3\beta_2 + 4\beta_3 + 5\beta_4 + 6\beta_5 + 4\beta_6 + 2\beta_7 + 3\beta_8$

F_4  $\theta = 2\beta_1 + 3\beta_2 + 4\beta_3 + 2\beta_4$

G_2  $\theta = 2\beta_1 + 3\beta_2$

Denote by $E_i, F_i, i = 1, \ldots, N$ the Chevalley generators associated with $\mathcal{B}$: let $E_0 \in (\mathcal{G}_0)_{-\theta}$ and $F_0 \in (\mathcal{G}_0)_\theta$ be such that

$$[E_0, F_0] = -H_\theta \in h_0, \qquad (E_0, F_0)_0 = 1$$

where H_θ is defined by

$$(H_\theta, H) = \theta(H) \qquad \forall\, H \in h_0$$

Introduce the following elements of $\mathcal{G}$:

$$e_0 = t \otimes E_0, \quad f_0 = t^{-1} \otimes F_0, \quad h_0 = c - 1 \otimes H_\theta$$
$$e_i = 1 \otimes E_i, \quad f_i = 1 \otimes F_i, \quad h_i = 1 \otimes H_i, \qquad i = 1, \ldots, N$$

They generate a complex Lie subalgebra $\mathcal{L}_1$ of $\mathcal{G}$; let $\mathcal{L} = \mathcal{L}_1 \oplus \mathbf{C}d$ be another Lie subalgebra of $\mathcal{G}$. $\mathcal{L}$ has $h = h_0 \oplus \mathbf{C}c \oplus \mathbf{C}d$ as a Cartan subalgebra. If one defines $\delta \in h'$ by

$$\delta(h_i) = 0, \qquad i = 0, \ldots, N$$
$$\delta(d) = 1$$

Then $(\delta, \delta) = 0$. The root basis $\mathcal{B} = \{\alpha_0, \ldots, \alpha_N\}$ associated with $\mathfrak{h}$ is the following:

$$\alpha_i(h) = \beta_i(h), \quad h \in \mathfrak{h}_0, \qquad \alpha_i(c) = \alpha_i(d) = 0, \quad i = 1, \ldots, N.$$
$$\alpha_0(h) = -\theta(h), \quad h \in \mathfrak{h}_0, \qquad \alpha_0(c) = 0, \quad \alpha_0(d) = 1.$$

and one has

Theorem 4. (i) $\{e_i, f_i, h_i\}_{i,j=0,\ldots,N}$, together with the G.C.M. $A = \{2(\alpha_i, \alpha_j)/(\alpha_i, \alpha_i)\}_{i,j=0,\ldots,N}$ generate the Lie algebra $\mathcal{L} = \mathcal{G}(A)$, of the type $\mathbf{X}_N^{(1)}$.

(ii) This Lie algebra is isomorphic to the loop algebra $\mathcal{G}$.

(iii) δ is the smallest positive isotropic root of $\mathcal{G}$.

SKETCH OF THE PROOF [35]. (i) One easily sees that A is a G.C.M, and that the relations of Lemma 1 are satisfied. That $\mathcal{L}$ is of the type $\mathbf{X}_N^{(1)}$ is seen by inspection of the G.C.M. A.

(ii) It suffices to check (by induction on n) that any element $t^n \otimes x$ of $\mathcal{G}$ belongs to $\mathcal{L}$. □

Corollary. (i) Any affine Kac-Moody algebra of type $\mathbf{X}_N^{(1)}$ (see Table 1) can be realized as the loop algebra of $\mathbf{X}_N$.

(ii) The multiplicity of any isotropic root of $\mathbf{X}_N^{(1)}$ is equal to N.

The Cases k ≠ 1 (Twisted Cases)

Let $\mathcal{G}_0$ be a simple Lie algebra of the type $\mathbf{A}_N$, $\mathbf{D}_N$, or $\mathbf{E}_6$ (see Table 2). The Dynkin diagrams of these Lie algebras possess remarkable symmetry properties: they are invariant under the following transformations:

$$\mathbf{A}_N\!: \quad \beta_i \to \beta_{N+1-i}; \quad \beta_{N+1-i} \to \beta_i, \quad i = 1, \ldots, N$$

$$\mathbf{D}_N\!: \quad \beta_i \to \beta_i \quad (i = 1, \ldots, 1-2); \quad \beta_I \to \beta_{I+1}; \quad \beta_{I+1} \to \beta_I$$

$$\mathbf{D}_4\!: \quad \beta_1 \to \beta_3; \quad \beta_3 \to \beta_4; \quad \beta_4 \to \beta_1; \quad \beta_2 \to \beta_2$$

$$\mathbf{E}_6\!: \quad \beta_1 \to \beta_5; \quad \beta_5 \to \beta_1; \quad \beta_2 \to \beta_4; \quad \beta_4 \to \beta_2; \quad \beta_3 \to \beta_3; \quad \beta_6 \to \beta_6$$

Note that all these diagram automorphisms are of order 2, except the one concerning $\mathbf{D}_4$ (this is the celebrated *triality property* of $\mathbf{D}_4$) [16].

Denote by ρ the *straight extension* [17] of the diagram automorphism ρ_0 of $\mathcal{G}_0$, i.e., the Lie algebra automorphism defined by its action on the Chevalley generators:

$$\rho \cdot e_i = e_{\rho_0(i)}, \quad \rho \cdot f_i = f_{\rho_0(i)}, \quad i = 1, \ldots, N$$

With respect to the action of ρ, $\mathcal{G}_0$ has the following decomposition:

$$\mathcal{G}_0 = \sum_{p=0}^{k-1}{}^{\oplus}\, \mathcal{G}_{0;p}$$

where k is the order of ρ ($k = 2$ or 3), and

$$\mathcal{G}_{0;p} = \left\{ x \in \mathcal{G}_0 \text{ s.t. } \rho \cdot x = \exp\left(-2i\pi \frac{p}{k} \right) x \right\}$$

One can then form the Lie algebra

$$\mathcal{G} = \sum_{p \in \mathbf{Z}}{}^{\oplus}\, (t^p \otimes \mathcal{G}_{0;p \bmod k}) \oplus \mathbf{C}c \oplus \mathbf{C}d$$

as a Lie subalgebra of $\mathbf{C}[t, t^{-1}] \otimes \mathcal{G}_0 \oplus \mathbf{C}c \oplus \mathbf{C}d$. Denote by E_i, F_i, $i = 1, \ldots, N$, the Chevalley generators of $\mathcal{G}_0$ and let $H_i = [E_i, F_i]$. Introduce the following elements of $\mathcal{G}_{0;0}$:

(i) $\mathcal{G}_0 = \mathbf{A}_{2l}$

$$E_i = E_i + E_{2l+1-i}, \quad F_i' = F_i + F_{2l+1-i}, \quad H_i' = H_i + H_{2l+1-i}, \quad i = 1, \ldots, l-1$$

$$E_l' = 2(E_l + E_{l+1}), \quad F_l' = 2(F_l + F_{l+1}), \quad H_l' = 2(H_l + H_{l+1})$$

(ii) $\mathcal{G}_0 = \mathbf{A}_{2l-1}$

$$E_i' = E_i + E_{2l+1-i}, \quad F_i' = F_i + F_{2l+1-i}, \quad H_i' = H_i + H_{2l+1-i}, \quad i = 1, \ldots, l-1$$

$$E_l' = E_l, \quad F_l' = F_l, \quad H_l' = H_l$$

(iii) $\mathcal{G}_0 = \mathbf{D}_{l+1}$

$$E_i' = E_i, \quad F_i' = F_i, \quad H_i' = H_i, \qquad i = 1, \ldots, l-1$$

$$E_l' = E_l + E_{l+1}, \quad F_l' = F_l + F_{l+1}, \quad H_l' = H_l + H_{l+1}$$

(iv) $\mathcal{G}_0 = \mathbf{E}_6$

$$E_1' = E_1 + E_5, \quad F_1' = F_1 + F_5, \quad H_1' = H_1 + H_5$$

$$E_2' = E_2 + E_4, \quad F_2' = F_2 + F_4, \quad H_2' = H_2 + H_4$$

$$E_3' = E_3, \quad F_3' = F_3, \quad H_3' = H_3$$

$$E_4' = E_6, \quad F_4' = F_6, \quad H_4' = H_6$$

(v) $\mathcal{G}_0 = \mathbf{D}_4$

$$E_1' = E_1 + E_3 + E_4, \quad F_1' = F_1 + F_3 + F_4, \quad H_1' = H_1 + H_3 + H_4$$

$$E_2' = E_2, \quad F_2' = F_2, \quad H_2' = H_2$$

The following result is a standard consequence of the simple Lie algebra theory.

Lemma 5. (i) $\mathcal{G}_{0;0}$ is a simple Lie algebra, of the type $\mathbf{B}_l$, $\mathbf{C}_l$, $\mathbf{B}_l$, $\mathbf{F}_4$, and $\mathbf{G}_2$ respectively, with Chevalley generators E_i', F_i', H_i', $i = 1, \ldots, l$ ($l = 4$ for F_4, $l = 2$ for G_2).

(ii) $\mathcal{G}_{0;1}$ is an irreducible highest weight $\mathcal{G}_{0;0}$-module. Moreover, if one denotes by $\beta_1', \ldots, \beta_l'$ the roots of $\mathcal{G}_{0;0}$ attached to the Chevalley generators E_i', F_i', H_i', $i = 1, \ldots, l$, the expression of the highest weight θ' is given by

$\mathcal{G}_0 = \mathbf{A}_{2l}$: $\quad \theta' = 2(\beta_1' + \beta_2' + \cdots + \beta_l')$

$\mathcal{G}_0 = \mathbf{A}_{2l-1}$: $\quad \theta' = \beta_1' + 2(\beta_2' + \cdots + \beta_{l-1}') + \beta_l'$

$\mathcal{G}_0 = \mathbf{D}_{l+1}$: $\quad \theta' = \beta_1' + \beta_2' + \cdots + \beta_l'$

$\mathcal{G}_0 = \mathbf{E}_6$: $\quad \theta' = \beta_1' + \beta_2' + 3\beta_3' + 2\beta_4'$

$\mathcal{G}_0 = \mathbf{D}_4$: $\quad \theta' = \beta_1' + 2\beta_2'$

(iii) If $\mathcal{G}_0 = \mathbf{D}_4$ (then $k = 3$), $\mathcal{G}_{0;2}$ is an irreducible highest weight $\mathcal{G}_{0;0}$-module equivalent to $\mathcal{G}_{0;1}$.

Denote by $\mathfrak{h}_{0;0}$ the corresponding Cartan subalgebra. The rest of the discussion is identical to the untwisted case: choose $E'_0 \in (\mathcal{G}_{0;0})_{-\theta'}$, $F'_0 \in (\mathcal{G}_{0;0})_{\theta'}$ such that $[E'_0, F'_0] = -H_{\theta'}$ and $(E'_0, F'_0)_0 = 1$. Introduce the following elements of $\mathcal{G}$:

$$e_0 = t \otimes E'_0 \quad f_0 = t^{-1} \otimes F'_0 \quad h_0 = c - 1 \otimes H_\theta$$
$$e_i = 1 \otimes E'_i \quad f_i = 1 \otimes F'_i \quad h_i = 1 \otimes H'_i \qquad i = 1, \ldots, l$$

Set $\mathcal{L} = \mathcal{L}_1 \oplus \mathbf{C}d$, where $\mathcal{L}_1$ is the complex $\mathcal{G}$-subalgebra generated by $\{e_i, f_i, h_i\}_{i=0,\ldots,l}$. $\mathcal{L}$ has $\mathfrak{h} = \mathfrak{h}_{0;0} \oplus \mathbf{C}c \oplus \mathbf{C}d$ as a Cartan subalgebra, and if $\delta \in \mathfrak{h}'$ is defined by

$$\delta(h_i) = 0, \quad i = 0, \ldots, l \qquad \delta(d) = 1$$

Then $(\delta, \delta) = 0$.

The root basis $\mathcal{B} = \{\alpha_0, \ldots, \alpha_l\}$ associated to $\mathfrak{h}$ is characterized by

$$\alpha_i(h) = \beta'_i(h), \qquad h \in \mathfrak{h}_{0;0}, \qquad \alpha_i(c) = \alpha_i(d) = 0$$
$$\alpha_0(h) = -\theta'(h), \quad h \in \mathfrak{h}_{0;0}, \qquad \alpha_0(c) = 0, \quad \alpha_0(d) = 1$$

and one then has the counterpart of Theorem 4:

Theorem 5. (i) $\{e_i, f_i, h_i\}_{i=0,\ldots,l}$, together with the G.C.M. $A = \{2(\alpha_i, \alpha_j)/(\alpha_i, \alpha_i)\}_{i,j=0,\ldots,l}$ generate the Lie algebra $\mathcal{L} = \mathcal{G}(A)$, of the type $\mathbf{X}_N^{(k)}$.

(ii) This Lie algebra is isomorphic to the twisted loop algebra $\mathcal{G}$.

(iii) δ is the smallest positive isotropic root of $\mathcal{G}$.

The proof is similar to that of Theorem 4.

Corollary. (i) Any affine Kac-Moody algebra of type $\mathbf{X}_N^{(k)}$ can be realized as a twisted loop algebra of $\mathbf{X}_N$.

(ii)

$$\text{Dim } \mathcal{G}_{s\delta} = l \qquad \text{if } s \equiv 0 \bmod k$$

$$\text{Dim } \mathcal{G}_{s\delta} = \frac{N - l}{k - 1} \qquad \text{if } s \not\equiv 0 \bmod k$$

Some Remarks about Twist and Gradations

Let $\mathcal{G} = \mathcal{G}(A)$ be a Kac-Moody algebra of rank $n = l + 1$, and let $s = \{s_0, \ldots, s_l\}$ be a $(l + 1)$-tuple of integers. This allows us to endow $\mathcal{G}$

with the structure of **Z**-graded Lie algebra, the degree being defined by

$$\deg(e_i) = s_i, \quad \deg(f_i) = -s_i, \qquad i = 0, \ldots, l$$
$$\deg(\mathfrak{h}) = 0$$

This gradation is called the *gradation of type s*.

Consider for instance the loop algebra realizations of Section 5.2.2. The Lie algebra $\mathcal{G}$ constructed there has a natural gradation given by the adjoint action of the $d = t(d/dt)$ element. Clearly one has

$$[d, e_i] = [d, f_i] = 0, \quad i = 1, \ldots, l, \qquad [d, e_0] = -[d, f_0] = 1$$

This gradation is a gradation of type $\{1, 0, \ldots, 0\}$, usually called the *basic* or *homogeneous gradation*.

Actually, gradations of different types can naturally arise in different loop algebra realizations of $\mathcal{G}$. Let us illustrate this remark by the following simple example. Take $\mathcal{G}_0 = \mathbf{A}_2$, with the same notation as in the previous subsections for the Chevalley generators E_1, E_2, F_1, F_2; set $E_\theta = [E_1, E_2]$ and $F_\theta = [F_2, F_1]$. Let $w_0 = w_{\beta_2} \cdot w_{\beta_1}$ be the Coxeter element of $\mathcal{G}_0$ (i.e., a kind of "maximal" element in the Weyl group of $\mathcal{G}_0$). Let w be the straight extension of w_0, i.e., w is defined by

$$w \cdot E_1 = F_\theta, \quad w \cdot F_1 = E_\theta, \quad w \cdot E_2 = E_1, \quad w \cdot F_2 = F_1.$$

w is of order 3 and induces the following decomposition of $\mathcal{G}_0$:

$$\mathcal{G}_0 = \sum_{p=0}^{2} \mathcal{G}_{0;p}$$
$$\mathcal{G}_{0;p} = \{x \in \mathcal{G}_0 \text{ s.t. } w \cdot x = e^{-2i\pi/3}\, x\}$$

Set $\varepsilon = \exp(-2i\pi/3)$, and introduce the following elements of $\mathcal{G}_0$:

$$H'_1 = E_1 + F_\theta + E_2 \quad H'_2 = F_1 + E_\theta + F_2$$
$$A_1 = H_1 - \varepsilon^2 H_2, \quad A_2 = E_1 + \varepsilon^2 F_\theta + \varepsilon E_2, \quad A_3 = F_1 + \varepsilon^2 E_\theta + F_2.$$
$$B_1 = H_1 - \varepsilon H_2, \quad B_2 = E_1 + \varepsilon F_\theta + \varepsilon^2 E_2, \quad B_3 = F_1 + \varepsilon E_\theta + \varepsilon^2 F_2.$$

Then one easily sees that

$$\mathcal{G}_{0;0} = \mathbf{C}H'_1 \oplus \mathbf{C}H'_2$$
$$\mathcal{G}_{0;1} = \mathbf{C}A_1 \oplus \mathbf{C}A_2 \oplus \mathbf{C}A_3$$
$$\mathcal{G}_{0;2} = \mathbf{C}B_1 \oplus \mathbf{C}B_2 \oplus \mathbf{C}B_3$$

REMARK. $\mathcal{G}_{0;0}$ is a Cartan subalgebra of $\mathcal{G}_0$. One can then diagonalize the adjoint representation of $\mathcal{G}_0$ with respect to $\mathcal{G}_{0;0}$ in the following way: set

$$E'_1 = \tfrac{1}{3}[A_1 + A_2 + \varepsilon A_3] \quad F'_1 = \tfrac{1}{3}[B_1 + \varepsilon^2 B_2 + B_3]$$
$$E'_2 = \tfrac{1}{3}[A_1 + \varepsilon A_2 + A_3] \quad F'_2 = \tfrac{1}{3}[B_1 + B_2 + B_3]$$
$$E'_0 = \tfrac{1}{3}[\varepsilon A_1 + A_2 + A_3] \quad F'_1 = \tfrac{1}{3}[\varepsilon^2 B_1 + B_2 + B_3]$$
$$H''_1 = \frac{1}{\varepsilon - 1}[H'_1 - \varepsilon H'_2] \quad H''_2 = \frac{1}{\varepsilon - 1}[\varepsilon H'_1 - H'_2] \quad H''_0 - H''_2$$

and introduce the w-twisted loop algebra associated to $\mathcal{G}_0$:

$$\mathcal{G} = \sum_{p\in\mathbf{Z}}^{\oplus} \mathbf{C}t^p \otimes \mathcal{G}_{0;p\bmod 3} \oplus \mathbf{C}c \oplus \mathbf{C}d$$

provided with the standard Lie bracket and Killing form.

A long but easy computation then shows that the following elements of $\mathcal{G}$

$$e_i = t \otimes E'_i, \quad f_i = t^{-1} \otimes F'_i, \quad h_i = c + 1 \otimes H''_i, \qquad i = 0, \ldots, 2$$

are the Chevalley generators of the affine Kac-Moody algebra $\mathcal{G}(A)$ associated to the G.C.M.

$$A = \begin{pmatrix} 2 & -1 & -1 \\ -1 & 2 & -1 \\ -1 & -1 & 2 \end{pmatrix}$$

and then $\mathcal{G} \cong \mathcal{G}(A) = \mathbf{A}_2^{(1)}$.

The adjoint action of the d-element naturally defines on $\mathcal{G}$ a $\mathbf{Z}$-gradation

$$\deg(e_i) = -\deg(f_i) = 1 \qquad i = 0, \ldots, 2$$

called the *principal gradation* [35].

More generally, let w be the straight extension of some element w_0 in the Weyl group of a semisimple Lie algebra $\mathcal{G}_0$. Construct as above the w-twisted loop algebra $\mathcal{G}$ associated to $\mathcal{G}_0$. Then $\mathcal{G}$ is isomorphic to the untwisted loop algebra $\mathcal{G}_0^{(1)}$, and is naturally provided with a $\mathbf{Z}$-gradation induced by w, a priori different from the homogeneous gradation. Similarly in the construction of Section 5.2.2, replacing the diagram automorphism, ρ_0 by the product of ρ_0 by some inner automorphism w_0 of the root system of $\mathcal{G}_0$ leads to a twisted loop algebra $\mathcal{G}$ isomorphic to $\mathcal{G}_0^{(k)}$, with a natural gradation a priori different from the homogeneous gradation appearing in the ρ-twisted case. This is the original construction of Kac and Moody, which led to the classification of finite order automorphisms of simple finite-dimensional Lie algebras. The twisted realisations of affine Lie algebras will appear later, when discussing the vertex operator algebras.

5.2.3 Unitary Highest Weight Representations

Preliminaries

Let $\mathcal{G}$ be a complex Lie algebra, and assume that there is a consistent conjugation ω on $\mathcal{G}$ and a $\mathcal{G}$-subalgebra $\mathfrak{p}$ (*parabolic subalgebra*) such that

$$\mathfrak{p} + \omega \cdot \mathfrak{p} = \mathcal{G}.$$

Let $\pi: \mathcal{G} \to GL(\mathcal{V})$ be a linear representation of $\mathcal{G}$ in some linear space $\mathcal{V}$.

Definition 4. (i) π is a *highest weight representation* if there exists a character $\Lambda: \mathfrak{p} \to \mathbf{C}$ of $\mathfrak{p}$ and a vector $v_\Lambda \in \mathcal{V}$ such that

$$\pi[\mathcal{U}(\mathcal{G})] \cdot v_\Lambda = \mathcal{V},$$
$$\pi(b) \cdot v_\Lambda = \Lambda(b) v_\Lambda, \ \forall\, b \in \mathfrak{p}.$$

(ii) The highest weight presentation π of G is said to be *unitarizable* if there exists a positive-definite hermitian form $\mathcal{H}$ on $\mathcal{V}$ such that

$$\mathcal{H}(v_\Lambda, v_\Lambda) = 1.$$
$$\mathcal{H}(\pi(x) \cdot u, v) = \mathcal{H}(u, \pi(\omega \cdot x)v), \ \forall\, u, v \in \mathcal{V}, x \in \mathcal{G} \ (\textit{contravariance condition}).$$

REMARK. When restricted to the real Lie algebra $\mathcal{G}_\omega$, the contravariance condition yields

$$\mathcal{H}(\pi(x) \cdot u, v) = -\mathcal{H}(u, \pi(x) \cdot v) \qquad \forall\, u, v \in \mathcal{V}, \quad x \in \mathcal{G}_\omega,$$

leading then to a representation of $\mathcal{G}_\omega$ by antihermitian operators (corresponding to a unitary representation of the corresponding group, when π integrates to such a representation).

Verma Modules

Let $\mathfrak{p}$ be a parabolic $\mathcal{G}$-subalgebra, associated with the consistent conjugation ω. Let $\mathfrak{n}$ be a $\mathcal{G}$-subalgebra such that the following decomposition holds:

$$\mathcal{G} = \mathfrak{n} \oplus \mathfrak{p}.$$

This induces the following decomposition of the universal enveloping algebra $\mathcal{U}(\mathcal{G})$ of $\mathcal{G}$ [35, 7]:

$$\mathcal{U}(\mathcal{G}) = \mathfrak{n}\mathcal{U}(\mathcal{G}) \oplus \mathcal{U}(\mathfrak{p}).$$

If Λ is any complex character of $\mathfrak{p}$, denote by $I(\Lambda)$ the left $\mathcal{U}(\mathcal{G})$-ideal generated by the elements of the form $p - \Lambda(p)$, $p \in \mathfrak{p}$. The *Verma module* [71] associated with Λ is then defined by

$$M(\Lambda) = \mathcal{U}(\mathcal{G})/I(\Lambda)$$

Clearly, any highest weight $\mathcal{G}$-module with highest weight Λ is a quotient of $M(\Lambda)$ (see, e.g., [35, 9, 71]). Denote by v_Λ the canonical image of the unit $1 \in \mathcal{U}(\mathcal{G})$ in this quotient. $M(\Lambda)$ is obviously a highest $\mathcal{G}$-module, with highest weight Λ and highest weight vector v_Λ.

Extend Λ to a representation of $\mathcal{U}(\mathfrak{p})$, and denote by P the projection of $\mathcal{U}(\mathcal{G})$ onto $\mathcal{U}(\mathfrak{p})$, parallel to $\mathfrak{n}\mathcal{U}(\mathcal{G})$. Assuming that $\Lambda[P(\omega \cdot v)] = \Lambda[P(v)]^*$, $\forall\, v \in M(\Lambda)$, one easily checks that the bilinear form defined by

$$\mathcal{H}_1(u, v) = \Lambda[P(\omega \cdot v)u], \qquad u, v \in M(\Lambda)$$

is a contravariant Hermitian form on $M(\Lambda)$. Finally, let $R(\Lambda)$ be the radical of $\mathcal{H}_1$ on $M(\Lambda)$ and set

$$L(\Lambda) = M(\Lambda)/R(\Lambda)$$

$\mathcal{H}_1$ induces on $L(\Lambda)$ a nondegenerate contravariant Hermitian form $\mathcal{H}$. In the following, we will study $L(\Lambda)$ (let us remark that there is a one to one correspondence between the set of $L(\Lambda)$ modules and the dual $\mathfrak{h}'$ of $\mathfrak{h}$).

Clearly, the canonical action of the Cartan subalgebra $\mathfrak{h}$ of $\mathcal{G}$ on $L(\Lambda)$ can be diagonalized. The spectrum of this action is called the *weight system* of the representation. One then writes

$$L(\Lambda) = \sum_{\lambda \in \mathfrak{h}'}^{\oplus} L(\Lambda)_\lambda$$

$$L(\Lambda)_\lambda = \{v \in L(\Lambda) \text{ s.t. } h \cdot v = \lambda(h)v \ \forall\, h \in \mathfrak{h}\}$$

Let $\mathcal{G} = \mathcal{G}(A)$ be an affine Kac-Moody algebra. It is convenient to introduce the following lattices:

The *weight lattice*: $\mathcal{P} = \{\lambda \in \mathfrak{h}' \text{ s.t. } \lambda(h_i) \in \mathbf{Z},\ i = 0, \ldots, l\}$.

The *integral weight lattice*: $\mathcal{P}_+ = \{\lambda \in \mathcal{P} \text{ s.t. } \lambda(h_i) \geq 0,\ i = 0, \ldots, l\}$.

We note that $\mathcal{P}$ is the dual of the root lattice Q, with respect to the Killing form.

To investigate the unitary highest weight representations, let us start with an analysis of the possible parabolic subalgebras $\mathfrak{p}$.

Standard, Natural, and Mixed Sets of Positive Roots

The conjugation ω being assumed consistent (i.e., $\omega \cdot \mathcal{G}_\alpha = \mathcal{G}_{-\alpha}$, $\forall\, \alpha \in \mathcal{R}$), the first step in this analysis is to find all the possible sets of positive roots in $\mathcal{R}$. Assume that $\mathcal{G}$ is semisimple and finite-dimensional. If $\mathcal{B} = \{\alpha_1, \ldots, \alpha_l\}$ is any basis of simple roots of $\mathcal{G}$, define the standard set of positive roots by

$$\mathcal{R}_+^{\mathrm{st}} = \left\{\alpha \in \mathcal{R},\ \alpha = \sum_{i=1}^{l} k_i\alpha_i,\ k_i \in \mathbf{Z}^+\right\}$$

It is then a well-known result that any set of positive roots is Weyl-conjugate to $\mathcal{R}_+^{\mathrm{st}}$. Then taking for $\mathfrak{p}$ any Borel subalgebra of the form:

$$\mathfrak{b} = \mathfrak{h} \oplus \sum_{\alpha\in\mathcal{R}_+}^{\oplus} \mathcal{G}_\alpha$$

leads to equivalent representations. This result has recently been used for the classification of unitary highest weight representations of finite-dimensional semisimple Lie algebras [28].

From now on, assume that $\mathcal{G}$ is an affine Kac-Moody algebra, and let $\mathcal{B} = \{\alpha_0, \ldots, \alpha_l\}$ be the basis of simple roots introduced in Section 5.2.1. The standard set of simple roots can also be defined by

$$\mathcal{R}_+^{\mathrm{st}} = \left\{\alpha \in \mathcal{R},\ \alpha = \sum_{i=1}^{l} k_i\alpha_i,\ k_i \in \mathbf{Z}^+\right\}$$

but any set of positive roots is no longer Weyl-conjugate to $\mathcal{R}_+^{\mathrm{st}}$. Actually, in the $\mathbf{A}_1^{(1)}$ case, there are only two nonequivalent sets of positive roots: the standard and the natural ones, given by

$$\mathcal{R}_+^{\mathrm{st}} = \{\gamma + n\delta,\ \gamma \in \mathcal{R}_{\mathfrak{f}},\ n \in \mathbb{N}\} \cup \{\gamma \in (\mathcal{R}_{\mathfrak{f}})_+\}$$

respectively

$$\mathcal{R}_+^{\mathrm{nat}} = \{\gamma + n\delta,\ \gamma \in (\mathcal{R}_{\mathfrak{f}})_+,\ n \in \mathbf{Z}\} \cup \{n\delta,\ n \in \mathbf{Z}\}$$

The general case is more complex. Let $\mathcal{S}$ be any subset of the gradient root system $\mathcal{R}_{\mathfrak{f}}$. One can then define the mixed set of positive roots associated with $\mathcal{S}$ (if $\mathcal{S} \neq \mathcal{R}_{\mathfrak{f}}$).

If $\mathcal{R} \neq \mathbf{A}_{2l}^{(2)}$, we define

$$\begin{aligned}\mathcal{R}_+^{\mathcal{S}} = {} & \{\alpha + nk_\alpha\delta,\ \alpha \in \mathcal{R}_{\mathfrak{f}}^+ \setminus (\mathbf{Z}\mathcal{S} \cap \mathcal{R}_{\mathfrak{f}}^+,\ n \in \mathbf{Z}\} \\ & \cup\{\alpha + nk_\alpha\delta,\ \alpha \in \mathbf{Z}\mathcal{S} \cap \mathcal{R}_{\mathfrak{f}},\ n \in \mathbb{N}^*\} \\ & \cup\{\alpha \in \mathbf{Z}\mathcal{S} \cap \mathcal{R}_{\mathfrak{f}}^+\} \cup \{n\delta,\ n \in \mathbf{Z}\}\end{aligned}$$

(where k_α always equals 1 if $\mathcal{G}$ is of the type $\mathbf{X}_N^{(1)}$ or $\mathcal{G}$ is of the type $\mathbf{X}_N^{(k)}$ and α is a long root), and the corresponding one if $\mathcal{R} = \mathbf{A}_{2l}^{(2)}$. It is then straightforward to see that any set of positive roots is Weyl-conjugate to a $\mathcal{R}_+^{\mathcal{S}}$, for some subset $\mathcal{S}$ of $\mathcal{R}_{\mathfrak{l}}$ (see [29] for a more detailed analysis). To such a $\mathcal{R}_+^{\mathcal{S}}$, we associate the corresponding Borel subalgebra:

$$\mathfrak{b}^{\mathcal{S}} = \mathfrak{h} \oplus \sum_{\alpha \in \mathcal{R}+}^{\oplus} \mathcal{G}_\alpha$$

and build the corresponding $\mathcal{G}$-module $L(\Lambda)$, which we sometimes denote by $L_{\mathfrak{b},\omega}^{\mathcal{S}}(\Lambda)$ to emphasize the dependence on $\mathcal{S}$ and ω.

It is convenient to start the analysis with the case of the standard set of positive roots, leading to the so-called integrable representations.

Integrable Representations

Take $\mathcal{S} = \mathcal{R}_{\mathfrak{l}}$, and set $\mathfrak{b}^{\text{st}} = \mathfrak{b}^{\mathcal{R}_{\mathfrak{l}}}$. Then we have

Proposition 3. In the case where $\mathfrak{p} = \mathfrak{b}^{\text{st}}$, only the Cartan conjugation can lead to unitarizable representations of $\mathcal{G}$ (and thus to unitary representations of the compact real form $\mathcal{G}_c$ of $\mathcal{G}$).

PROOF. The first step is the inspection of the $\mathbf{A}_l^{(1)}$ case; for simplicity, it is convenient to work with the (untwisted) loop algebra realization. Let then $\{e, f, h\}$ be a set of Chevalley generators of $s1(2, \mathbf{C})$, and set

$$e^n = t^n \otimes c, \quad f^n = t^n \otimes f, \quad h^n = t^n \otimes h$$

Assume that π is a highest weight representation of $\mathbf{A}_l^{(1)}$ on $\mathcal{V} = L_{\mathfrak{b},\omega}^{\mathcal{S}}(\Lambda)$. with the highest weight Λ and highest weight vector v_Λ. Denote by $\mathcal{H}$ the contravariant Hermitian form on $\mathcal{V}$. The contravariance of $\mathcal{H}$ implies that

$$\pi(h^n) \cdot v_\Lambda = \Lambda(h^n) v_\Lambda = \Lambda_0 \delta_{n,0} v_\Lambda \qquad \textit{with } \Lambda_0 \in \mathcal{R}$$

The conjugation ω is completely determined by $\omega \cdot e = s_e f$; $\omega \cdot h^1 = s_h h^{-1}$ with $s_e, s_h \in \{\pm 1\}$ (since ω is consistent). Let $j \in \mathbb{N}$ be such that $\pi(e^{-j}) \cdot v_\Lambda \neq 0$ (such a j always exists). Then

$$\mathcal{H}(\pi(f^{-n})\pi(e^{-j})^k \cdot v_\Lambda, \pi(f^{-n})\pi(e^{-j})^k \cdot v_\Lambda)$$
$$= s_h^n s_e [\Lambda_0 + n\pi(c) + 2k] \mathcal{H}(\pi(e^{-j})^k \cdot v_\Lambda, \pi(e^{-j})^k \cdot v_\Lambda)$$

for any k, n implies that $s_e = s_h = 1$. Applying this procedure to any $\mathbf{A}_l^{(1)}$ subalgebra of $\mathcal{G}$ yields the result. □

Let $e \in \mathcal{G}$, and set $f = \omega_c \cdot e$, $h = [e, f]$, so that e, f and h span a $sl(2, \mathbf{C})$ $\mathcal{G}$-subalgebra. If π is a unitarizable highest weight representation of $\mathcal{G}$ with highest weight Λ and highest weight vector v_Λ, then we have

$$\|\pi(f)^n \cdot v_\Lambda\|^2 = n[\Lambda(h) - (n - 1)]\|\pi(f)^{n-1} \cdot v_\Lambda\|^2$$

Since π is unitarizable, the requirement of positive definiteness of the contravariant Hermitian form imposes that $\pi(f)$ is nilpotent on v_Λ. More precisely, $\Lambda(h)$ has to be an integer number, and one has $\pi(f)^{\Lambda(h)+1} \cdot v_\Lambda = 0$.

Clearly, a necessary and sufficient condition for this relation to be satisfied is the following: π has to be of the nilpotent type, i.e., all generators e_i, f_i $(i = 0, \ldots, n - 1)$ act as locally nilpotent operators on the representation space $\mathcal{V}$ (we recall that an operator A is locally nilpotent on $\mathcal{V}$ if $\forall\, v \in \mathcal{V}$, there exists $n \in \mathbb{N}$ such that $A^n \cdot v = 0$). Moreover, by the Poincaré-Birkhoff-Witt theorem, it is sufficient to check that the f_i generators act as nilpotent operators on the highest weight vector v_Λ [35,66]. In the particular case of Kac-Moody algebras, the representations of the nilpotent type are also called *integrable representations*.

Return now to the $\mathcal{G}$-module $L_{b^{st},\omega_c}(\Lambda)$. By the above argument, it is integrable if and only if $\Lambda(h_i) \in \mathbb{N}$, $i = 0, \ldots, n-1$, or, which is equivalent, $\Lambda \in \mathcal{P}_+$. Conversely, using the Poincaré-Birkhoff-Witt theorem again, the $\mathcal{G}$-module $L_{b^{st},\omega_c}(\Lambda)$ with $\Lambda \in \mathcal{P}_+$ is shown to be unitarizable ($\mathcal{H}$ is positive definite), and irreducible [35]. Summarizing, we have

Theorem 6. The $\mathcal{G}$-modules $L_{b^{st},\omega_c}(\Lambda)$ with $\Lambda \in \mathcal{P}_+$ exhaust all possible irreducible unitarizable highest weight modules with respect to the standard set of positive roots (and with respect to the Cartan conjugation). Moreover, any highest weight $\mathcal{G}$-module splits as a direct sum of modules $L(\lambda)$, $\lambda \in \mathcal{P}_+$ [35].

REMARK. $L_{b^{st},\omega_c}(\Lambda)$ with $\Lambda \in \mathcal{P}_+$ is thus identified to the unique irreducible quotient of the Verma module $M(\Lambda)$.

One also has the following corollary.

Corollary. The *Harish-Chandra formula* holds, i.e.,

$$L(\Lambda) = M(\Lambda) \Big/ \left(\sum_{i=0}^{n-1}{}^{\oplus} f_i^{\Lambda(h_i)-1} \cdot M(\Lambda) \right) \qquad \text{for any } \Lambda \in \mathcal{P}_+.$$

Another immediate consequence is the following

Corollary. $\Lambda(c)$ is a nonnegative integer number; moreover, $L_{b^{st},\omega_c}(\Lambda)$ is nontrivial if and only if $\Lambda(c) \neq 0$.

It is now easy to describe the weight system of $L(\Lambda)$:

$$L(\Lambda) = \sum_{\lambda \in \mathcal{P}(\Lambda)}{}^{\oplus} L(\Lambda)_\lambda$$

$$\mathcal{P}(\Lambda) = \{\lambda \in \mathfrak{h}' \text{ s.t. } L(\Lambda)_\lambda \neq \{0\}\}$$

The following result is a straightforward consequence of $su(2)$ representation theory.

Proposition 4. Let $\lambda \in \mathcal{P}(\Lambda)$, and let α be a real root of $\mathcal{G}$.

(i) $2\langle\lambda, \alpha\rangle/\langle\alpha, \alpha\rangle \in \mathbf{Z}$; moreover there exist two nonnegative integer numbers n_- and n_+ such that $\lambda + n\alpha \in \mathcal{P}(\Lambda)$ iff $n \in [-n_-, n_+] \cap \mathbf{Z} \cdot$ n_+ and n_- are related by

$$n_- - n_+ = 2\frac{\langle\lambda, \alpha\rangle}{\langle\alpha, \alpha\rangle}$$

(ii) The function $n \to \mathrm{Mult}_{L(\Lambda)}(\lambda + n\alpha)$ is symmetric with respect to $n = \frac{1}{2}[n_+ - n_-]$, and increasing in $[-n_-, \frac{1}{2}(n_+ - n_-)]$.

Moreover, one has the following geometric description of the weight system $\mathcal{P}(\Lambda)$ of $L(\Lambda)$.

Proposition 5 [35]. (i) $\mathcal{P}(\Lambda)$ is the intersection of $(\Lambda + Q)$ with the convex hull of the orbit of the Weyl group W of $\mathcal{G}$ through Λ (recall that the root lattice Q is the lattice generated by the root system $\mathcal{R}$).

(ii) $\mathcal{P}(\Lambda)$ is the union of the W-orbits through elements $\lambda \in \mathcal{P}_+$ such that $\lambda \leq \Lambda$.

Elementary and Exceptional Representations

Consider now the case where the set of positive roots is "nonstandard" (i.e., it is of the mixed or the natural type), and let $\alpha \in \mathcal{R}\backslash\mathbf{Z} \cap \mathcal{R}$ (α is a "natural root"). Introduce the corresponding Chevalley generators $\{e, f, h\}$ and set, as in the "standard" case,

$$e^n = t^n \otimes e, \quad f^n = t^n \otimes f, \quad h^n = t^n \otimes h, \qquad n \in \mathbf{Z}$$

Let us restrict ourselves to the corresponding $\mathbf{A}_l^{(1)}$-submodule of $L_{b^{\mathcal{G}},\omega}(\Lambda)$. Then

$$\mathcal{H}(f^n \cdot v_\Lambda, f^n \cdot v_\Lambda) = (s_h)^n s_e[\Lambda(h) - nc], \qquad n \in \mathbf{Z}$$

This clearly implies that

$$s_h = 1; \quad \Lambda(c) = 0; \quad s_e\Lambda(h) \geq 0$$

Moreover, since

$$\left\| \left[\sum_n u_n t^n\right] \otimes f \cdot v_\Lambda \right\|^2 = s_e\Lambda\left[\left(\sum_n u_n t^n\right)\left(\sum_n u_n^* t^{-n}\right) \otimes h\right]$$

for any $\Sigma_n u_n t^n \in \mathbf{C}[t, t^{-1}]$, it follows that $s_e\Lambda$ must be a positive definite linear functional on $\mathbf{C}[t, t^{-1}]$, which can be extended to a positive Radon measure m on the circle S^1.

Consider first the compact case, i.e., the case where $s_e = 1$. By standard $su(2)$ representation theory, there exists a proper nontrivial invariant submodule L' of $L(\Lambda)$ such that for any integral number $n > \Lambda(h)$, $f^n \cdot v_\Lambda \in L'$ (the nondegeneracy of $\mathcal{H}$ on L' can easily be checked, and L' is then nonempty).

$L_{\mathfrak{h}^{\text{nat}},\omega_c}(\Lambda)$ is then reducible.

Let us now state without proof the following result of [29].

Proposition 6 (Jakobsen-Kac). The $\mathbf{A}_1^{(1)}$-module $L_{\mathfrak{h}^{\text{nat}},\omega}(\Lambda)$ is irreducible if and only if the positive Radon measure m has infinite support.

As an immediate consequence, the Radon measure involved in the $\mathbf{A}_1^{(1)}$-module $L_{\mathfrak{h}^{\text{nat}},\omega}(\Lambda)$ is concentrated in a finite number of points. It is then easy to see that the corresponding representation is located, i.e., it can be constructed as follows: let $\varphi_1, \ldots, \varphi_N$ be a set of evaluation functionals

$$\varphi_i(t^k \otimes g) = c_i^k g, \qquad c_i \in \mathbf{C}$$

and let $\pi_1, \ldots, \pi_N$ be a set of unitarizable representations of $\mathbf{A}_1$, with respective highest weights $\lambda_1, \ldots, \lambda_N$. Then

$$\Pi_{(\{\varphi_i\},\{\pi_i\})} = (\pi_1 \circ \varphi_1) \oplus (\pi_2 \circ \varphi_2) \oplus \cdots \oplus (\pi_N \circ \varphi_N)$$

acting on

$$\mathcal{V}_{(\{\varphi_i\},\{\pi_i\})} = L(\lambda_1) \otimes L(\lambda_2) \otimes \cdots \otimes L(\lambda_N)$$

clearly defines a unitarizable highest weight representation of $\mathbf{A}_1^{(1)}$.

Let us return now to the case of a generic untwisted loop algebra $\mathcal{G}$, built over the simple Lie algebra $\mathcal{G}_0$ (complex). Let $\mathcal{G}_\omega$ be a real form corresponding to a Hermitian symmetric space,and let ω be the conjugation of $\mathcal{G}_0$ with respect to $\mathcal{G}_\omega$. Consider a Cartan decomposition of $\mathcal{G}_\omega$: $\mathcal{G}_\omega = \mathfrak{t}_\omega \oplus \mathfrak{p}_\omega$. Let $\mathfrak{h}_\omega$ be a maximal abelian subalgebra of $\mathfrak{t}_\omega$. Then the complexified $(\mathfrak{h}_\omega)^{\mathbf{C}}$ of $\mathfrak{h}_\omega$ is a Cartan subalgebra of $\mathcal{G}_0$. By general results on Cartan decompositions (see [26] for instance), $\mathfrak{t}_\omega$ is reductive in $\mathcal{G}_\omega$; it has a one-dimensional center $\mathcal{C}_\omega$, and one has $\mathfrak{t}_\omega = \mathfrak{t}'_\omega \oplus \mathcal{C}_\omega$, $\mathfrak{t}'_\omega = [\mathfrak{t}_\omega, \mathfrak{t}_\omega]$.

Moreover, $\mathcal{C}_\omega = \mathbf{R}c_\omega$, c_ω being chosen in such a way that its eigenvalues under its adjoint action on $(\mathfrak{p}_\omega)^{\mathbf{C}}$ are $\pm i$. Let

$$\mathfrak{p}^{\pm} = \{g \in (\mathfrak{p}_\omega)^{\mathbf{C}} \text{ s.t. } [c_\omega, g] = \pm ig\}$$

Let $\mathcal{R}_0$ be the root system of $\mathcal{G}_0$ associated to $(\mathfrak{h}_\omega)^{\mathbf{C}}$, and denote by $\mathcal{R}_t$ and $\mathcal{R}_p$ the subsets of compact and non-compact roots: $\mathcal{R} = \mathcal{R}_t \cup \mathcal{R}_p$.

It is then possible to choose a set of positive roots of $\mathcal{R}$ such that

$$\mathfrak{p}^+ = \sum_{\alpha \in \mathcal{R}_p^+} \mathcal{G}_\alpha$$

From the analysis of the compact case, it follows that $\mathcal{U}(\mathbf{C}[t, t^{-1}] \otimes \mathfrak{t}'_\omega) \cdot v_\Lambda$ is in the kernel of the contravariant Hermitian form $\mathcal{H}$. Moreover, by assumption:

$$\mathcal{U}(\mathbf{C}[t, t^{-1}] \otimes \mathfrak{p}^+) \cdot v_\Lambda = 0,$$

$$L(\Lambda) = \mathcal{U}(\mathbf{C}[t, t^{-1}] \otimes \mathfrak{p}^-) \cdot v_\Lambda,$$

and there is a positive Radon measure m on S^1 such that

$$(t^k \otimes c_\omega) \cdot v_\Lambda = -\int_{S^1} e^{ik\tau}\, dm(\tau).$$

In particular, $\Lambda(1 \otimes c_\omega) = -\int_{S^1} dm(\tau) < 0$; one then use the following result [70].

Lemma 6 (Vergne, Rossi) In the case where the real rank is greater than one, there exists a finite number $\lambda_1, \ldots, \lambda_p$ of integral values of $\Lambda(c_\omega)$ for which the module $\mathcal{U}(\mathfrak{p}^-). v_\Lambda$ is reducible.

As a direct consequence, and since it is easy to deform the m measure to one yielding one of those discrete values without changing the reducibility or irreducibility of $L(\Lambda)$, one is led to restrict to $\mathcal{G}_\omega = su(n, 1)$.

The corresponding representations of $\mathbf{C}[t, t^{-1}] \otimes su(n, 1)$ can be shown to be unitary [29], and are called the *exceptional representations.* To complete the analysis, there only remains to consider the case of twisted loop algebras. But since such algebras contain a "maximal" untwisted loop subalgebra different than one of the type $\mathbf{A}_n$, there is no room for exceptional representations.

Summarizing, one gets

Theorem 7. The exceptional representations of $\mathbf{C}[t, t^{-1}] \otimes su(n, 1)$ and the elementary representations of any real affine Lie algebra exhaust all unitary highest weight modules based on orderings of affine root systems different than the standard ordering.

Remarks on Kac-Moody Groups and Their Unitary Highest Weight Representations

In the previous sections, we discussed unitary highest weight representations of some Kac-Moody Lie algebras, namely the affine ones, realizable as central extensions of (untwisted or twisted) loop algebras. It is then relevant to try to construct the corresponding representations of the Kac-Moody groups. Let us stress that this remark does not intend to give a precise and complete account of the theory. Our aim here is only to give a general outlook of the relations between current groups and current algebras, and between their respective representation theories. For more details, we refer to [160] and [165], and references therein. Firstly, it must be said that, since we deal with infinite-dimensional Lie groups and Lie algebras, one can be led to introduce not only one, but several groups associated with a given Lie algebra, which are basically different completions of a smallest one. Different ways of constructing such groups can be used, and in his review paper [64] (to which we refer for more details), Tits is then led to separate them into three general procedures, namely formal, algebraic and analytic constructions.

We will here essentially describe the idea of the formal and algebraic aspects of Kac-Moody groups, since it allows to have a closer connection with the representation theory of Lie algebras. In the finite-dimensional case, Ado's theorem allows to imbed a Lie algebra $\mathcal{G}$ in the endomorphism group End(V) of some vector space V. It is then possible to introduce the exponential of any element of $\mathcal{G}$, and thus consider the group generated by exp($\mathcal{G}$), which admits $\mathcal{G}$ as a Lie algebra. In the infinite-dimensional case, since the $\mathcal{G}$-modules are infinite-dimensional, such a construction is no longer possible, and it is necessary to refine the method, for instance as follows, by means of the representation theory of the Lie algebra. Start from an affine Lie algebra $\mathcal{G}$, with Chevalley generators $\{e_i, f_i, h_i; i = 1, \ldots, n\}$, and consider a unitary $\mathcal{G}$-module V, on which $\mathcal{G}$ acts by the representation π, assuming that the representative of the Chevalley generators e_i and f_i, $i = 1, \ldots, n$ are locally nilpotent. The Kac-Moody group will then be identified with the End(V)-subgroup generated by $\exp(\mathbf{C}\pi(e_i))$ and $\exp(\mathbf{C}\pi(f_i))$. Moody [51] applied this method, starting from the adjoint representation of the Kac-Moody algebra, and Garland [19] (only in the affine case) and Marcuson [47] used highest weight representations. Unfortunately, the construction highly depends on the representation one starts with, and it is not known how the group finally obtained depends on this representation. Kac and Peterson [37] developed an alternative approach by considering simultaneously all the integrable representations of the Kac-Moody algebra.

Another possible approach to the construction of Kac-Moody groups is to try to build the group by generators and relations. Indeed, Tits [65] developed such an approach, but his attempt was not completely conclusive.

The analytic approach was essentially developed by quantum field theorists; the problem in this approach is to look at the Kac-Moody group as a central extension of some loop group. Then, one must give a construction of the two-cocycle describing the central extension. Various constructions have actually been proposed, and we refer to [13] for a description of the corresponding results. Let us however quote the work by Mickelsson [49], who built explicitely the central extension of $C(S^1, G)$ from a trivial 2-cocycle of $C(D, G)$, D being the unit disk, and also the work of Murray [53], who gave a construction of it from a group of paths over loop groups. Another construction of such a two-cocycle can also be found in a paper by Vershik, Gelfand and Graev [72], in which they consider a projective deformation of the energy representation. Actually, such discussions, viewed from the point of view of quantum field theory, opened the problem of "realistic" gauge groups based on more than one-dimensional space-time manifolds. For instance, Singer [59] proposed a geometric construction of the group and its actions. The particular case where the space-time manifold is a torus will be discussed in Section 5.3.

At the level of representation theory, one can quote the following results: Garland showed that the integrable representations of affine Kac-Moody algebras actually integrate to unitary representations of the corresponding group; Mickelsson's construction of the two-cocycle thus provides an explicit realisation of them, closely connected to the so-called Wess-Zumino-Witten model. Exceptional representations (in the case of $su(n, 1)$), like continuous tensor products representations (which exist for the same group, as discussed in the first chapter), are completely determined by a positive Radon measure on the circle S^1; in that sense, they seem similar, and the equivalence was proved by Delorme [8]; continuous tensor product representations are then highest weight representations in the case $G = SU(n, 1)$. In the case of $SO(n, 1)$, the continuous tensor product representations are not of highest weight type (this is an obvious corollary of Theorems 6 and 7); however, they do not seem to have known realisations at the Lie algebra level.

Characters and Character Formulas

Before concluding this section, it is convenient to introduce the additional notion of character of a highest weight representation. Indeed, the characters permit to simplify considerably some arguments in representation theory. They also made possible to prove some old conjectures of number theory (combinatorial identities). Let us finally mention that they are the building blocks for the study of two-dimensional statistical models and conformal field theories in physics.

First, to any $\lambda \in \mathfrak{h}'$, there is associated its exponential, that is the function

$$e^{\lambda}: \mathfrak{h} \to \mathfrak{h}, \qquad h \to e^{\lambda(h)}$$

Now, let $\mathcal{V}$ be a highest weight module over the affine Lie algebra $\mathcal{G}$, and consider its decomposition with respect to the action of $\mathfrak{h}$:

$$\mathcal{V} = \sum_{\lambda \in \mathcal{P}(\mathcal{V})}^{\oplus}$$

$\mathcal{P}(\mathcal{V})$ being the weight system of $\mathcal{V}$.

The character of $\mathcal{V}$ is then defined by

$$\mathrm{ch}_{\mathcal{V}} = \sum_{\lambda \in \mathcal{P}(\mathcal{V})} \dim(\mathcal{V}_{\lambda}) e^{\lambda}$$

Using a Jantzen filtration [30], one can see that

$$\mathrm{ch}_{\mathcal{V}} = \sum_{\lambda \in \mathfrak{h}'} \mathrm{Mult}_{\mathcal{V}}(L(\lambda)) \mathrm{ch}_{L(\lambda)}$$

Specializing to the case of integrable representations, Kac [33] was able to given an explicit form for the character (*Kac's character formula*).

Theorem 8. Let $\mathcal{G}$ be an affine Kac-Moody algebra, with root system $\mathcal{R}$ and Weyl group W, and let $L(\Lambda)$ be an integrable $\mathcal{G}$-module. Then one has

$$\begin{aligned} \mathrm{ch}_{L(\Lambda)} &= \frac{\sum_{w \in W} \varepsilon(w) e^{w(\Lambda + \rho) - \rho}}{\prod_{\alpha \in \mathcal{R}_+} (1 - e^{-\alpha})^{\mathrm{Mult}(\alpha)}} \\ &= \frac{\sum_{w \in W} \varepsilon(w) e^{w(\Lambda + \rho) - \rho}}{\sum_{w \in W} \varepsilon(w) e^{w(\rho) - \rho}} \end{aligned}$$

where $\varepsilon(w) = (-1)^{\mathfrak{l}(w)}$, $\mathfrak{l}(w)$ being the length of w (i.e., the number of components of the form $w_{\alpha i}$).

REMARK. The second identity follows from the first one, and is known as *Weyl's denominator identity*. In the case of $\mathbf{A}_2^{(2)}$, setting $x = e^{-\alpha_0}$ and $y = e^{-\alpha_1}$, this identity is nothing but the famous *Jacobi triple product identity*:

$$\prod_{n=1}^{\infty} (1 - x^n y^n)(1 - x^{n-1} y^n)(1 - x^n y^{n-1}) = \sum_{n \in \mathbf{Z}} (-1)^n x^{m(m-1)/2} y^{m(m+1)/2}$$

The character formula can also be put in another form, involving classical theta functions (see [35, Chap. 13] or [52] for a review). To do this, let us first introduce the M lattice (see [35, Chap. 6] for a discussion of the definition and of some important properties of M):

If the G.C.M. of $\mathcal{G}$ is symmetric or $k > 1$, then $M = \overline{Q}$, the orthogonal projection of the root lattice Q on the complex-linear span $\mathfrak{h}'_{\mathfrak{f}}$ of the gradient root system.

Otherwise, M is the canonical image (under the pairing $\mathfrak{h} \leftrightarrow \mathfrak{h}'$) of the orthogonal projection $\overline{Q}^\vee$ of the coroot lattice $Q^\vee$ on its "gradient component" $\mathfrak{h}_{\mathfrak{f}}$.

Let $\lambda \in \mathfrak{h}'$, and let $\bar{\lambda}$ be its orthogonal projection on $\mathfrak{h}'_{\mathfrak{f}}$. Let $m = \lambda(c)$. Let us introduce the associated classical theta function:

$$\Theta_\lambda = e^{m\Lambda_0} \sum_{\mu \in M + \bar{\lambda}/m} e^{-\frac{1}{2}m|\mu|^2\delta + m\mu}$$

The character formula then takes the form

$$\mathrm{ch}_{L(\Lambda)} = e^{|\Lambda+\rho|^2/2(m+\check{h}) - |\rho|^2/2\check{h}} \frac{\sum_{w \in W_{\mathfrak{f}}} \varepsilon(w)\Theta_{w\cdot(\Lambda+\rho)}}{\sum_{w \in W_{\mathfrak{f}}} \varepsilon(w)\Theta_{w\cdot\rho}}$$

(We recall that ρ is the Weyl element, $\check{h} = \rho(c)$ the dual Coxeter number, and $W_{\mathfrak{f}}$ the gradient Weyl group). We also note that in the simplest case $\Lambda = \Lambda_0$, one has a simpler formula for the character

$$\mathrm{ch}_{L(\Lambda_0)} = \frac{\Theta_{\Lambda_0}}{\varphi(q)}, \qquad \varphi(q) = \prod_{i=1}^{\infty} (1 - q^{-i})$$

5.3 SOME CONNECTED PROBLEMS AND APPLICATIONS

5.3.1 The Virasoro Algebra

As stressed in the first chapters, loop groups inherit a canonical action of the group of diffeomorphisms of the circle. At the infinitesimal level, it is then relevant to study the Lie algebra of this diffeomorphism group and its subalgebras, for instance the Lie algebra of polynomial vector fields on the circle. It actually appears that its representation theory is considerably rich, and is remarkably correlated with Kac-Moody representation theory.

Structure and Representations

Introduce first the Lie algebra of complex polynomial vector fields on the circle, the so-called Witt algebra:

$$\underline{\mathrm{Witt}} = \sum_{n\in\mathbf{Z}}^{\oplus} \mathbf{C}L_n, \qquad L_n = -t^{n+1}\frac{d}{dt}$$

which satisfies the relations $[L_m, L_n] = (m - n)L_{m+n}$. This algebra admits a triangular decomposition

$$\underline{\mathrm{Witt}} = \mathcal{L}_- \oplus \mathcal{W}_0 \oplus \mathcal{L}_+$$

$$\mathcal{L}_\pm = \sum_{\pm n>0}^{\oplus} \mathbf{C}L_n$$

$$\mathcal{W}_0 = \mathbf{C}L_0$$

The study of the complex cohomology of the Witt algebra was started by Gelfand and Goncharova [24], who showed that for $k > 1$, $H^k(\mathcal{L}_+, \mathbf{C})$ splits into two one-dimensional eigenspaces of L_0, with eigenvalues the opposites of the Euler pentagonal numbers (i.e., $(3k^2 \pm k)/2$). These numbers play an important role in the unitary highest weight representation theory of the Witt and Virasoro algebras [56]. Concerning the central extensions, it is easy to check that dim Z^2 (Witt, $\mathbf{C}$) = 2 and dim $H^2(\underline{\mathrm{Witt}}, \mathbf{C}) = 1$, and that the universal central extension of $\underline{\mathrm{Witt}}$, *the Virasoro algebra*, can be put in the form:

$$\underline{\mathrm{Vir}} = \underline{\mathrm{Witt}} \oplus \mathbf{C}\tilde{c}$$

$$[L_m, L_n] = (m - n)L_{m+n} + \frac{m^3 - m}{12}\tilde{c}\delta_{m+n,0}$$

$$[L_n, \tilde{c}] = 0$$

the form $(m^3 - m)\tilde{c}/12$ being purely conventional (inherited from string theory). $\underline{\mathrm{Vir}}$ also admits a triangular decomposition

$$\underline{\mathrm{Vir}} = \mathcal{L}_- \oplus \mathcal{L}_0 \oplus \mathcal{L}_+$$

with $\mathcal{L}_0 = \mathcal{W}_0 \oplus \mathbf{C}\tilde{c}$.

It is then relevant to look at the irreducible unitarizable highest weight $\underline{\mathrm{Vir}}$-modules, i.e., modules $\mathcal{V}$ on which the action of L_0 can be diagonalized:

$$\mathcal{V} = \sum_{\lambda\in\mathcal{L}_0'} \mathcal{V}_\lambda$$

such that:

(i) There exists a "vacuum vector" (highest weight vector) v:

$$L_j \cdot v = 0 \quad \text{if } j > 0, \qquad L_0 \cdot v = Ev$$

(ii) (L_0 is sometimes called the energy operator; E is the vacuum energy.) There exists a positive definite Hermitian form $\mathcal{H}$ on $\mathcal{V}$, associated to the conjugation of $\underline{\text{Vir}}$: $\omega \cdot L_n = L_{-n}$.

(iii) $\tilde{c}$ acts as $z \cdot 1$.

Actually, there always exists a nondegenerate contravariant form on $\mathcal{V}$; the question is rather whether this form is positive.

As in the Kac-Moody case, let us build the Verma module, $M(E, z)$, for given values of E and z. With respect to the action of L_0, $M(E, z)$ splits as

$$M(E, z) = \sum_{n \geq 0}^{\oplus} M(E, z)_{E+n}$$

(with the usual conventions).

Due to the contravariance of $\mathcal{H}$, it is clear that for $n \neq n'$, we have

$$M(E, z)_{E+n} \text{ is orthogonal to } M(E, z)_{E+n'}$$

As in the case of integrable representations of Kac-Moody algebras for instance, the radical of $\mathcal{H}$ is the unique maximal submodule of $M(E, z)$, and then unitarizability holds if and only if $\mathcal{H}$ is positive semidefinite on $M(E, z)$. Kac gave the following remarkable formula (*Kac's determinantal formula* [34], conjectured from Shapovalov's formula [58]), proved later on by Feigin and Fuchs [12].

Lemma 7. The determinant of the contravariant Hermitian form is given by

$$\text{Det } \mathcal{H}_{M(E,z)_{E+n}} = \prod_{j=1}^{n} \prod_{\substack{rs=j \\ s\leq r}} [E - E_{r,s}(z)]^{\varphi(E-j)}$$

with

$$E_{r,s}(z) = \tfrac{1}{48}[(13 - z)(r^2 + s^2) + \sqrt{(z-1)(z-25)}\,(r^2 - s^2) - 24rs - 2 + 2z]$$

and

$$\varphi(q) = \prod_{j\in\mathbf{Z}_+} (1 - q^{-j}).$$

Examinating the determinantal formula, it is possible to find the possible places of unitarity in the (E, z) plane. Friedan, Qiu, and Shenker

[18] did it in the context of two-dimensional conformal field theory. Their result can be summarized as follows.

Proposition 7. (i) Unitarity implies $z \geq 0$ and $E \geq 0$

(ii) For $z > 1$, the Verma module is unitarizable for any $E \geq 0$, and irreducible.

(iii) For $z = 1$, the Verma module $M(E, z)$ is unitarizable for any $E \geq 0$; it is degenerate for E-values of the form $E = m^2/4$, $m \in \mathbb{N}$.

(iv) For $z < 1$, the places of unitarizability correspond to the pairs (E, z) of the form

$$z_m = 1 - \frac{6}{(m+2)(m+3)}$$

$$E_{r,s}(z_m) = \frac{[(m+3)r - (m+2)s]^2 - 1}{4(m+2)(m+3)}, \qquad 1 \leq s \leq r \leq m+1$$

and correspond to degenerate Verma modules (*discrete series*).

SKETCH OF THE PROOF. The proof of the proposition is essentially based on a careful (graphical and theoretical) analysis of the sign and the zeros of the $E_{r,s}$ curves (see [18]). (i) is trivial from the $E_{r,s}$ plot.

(ii) follows from the fact that $E - E_{r,s}(z)$ does not vanish for $z > 1$, and is positive for a sufficiently large E.

(iii) is partially a consequence of (ii); the second part is derived by replacing z by 1 in the expression of $E_{r,s}$.

(iv) is the main result of [18] (see also [12]). A first observation of the $E_{r,s}$ curves exhibited there shows that for a given value of z, the determinant will acquire negative factors unless $E = E_{r,s}(z)$ for some numbers r, s. This shows that the Verma module must be degenerate in this case. □

As in the case of Kac-Moody algebras, the characters of the $\underline{\text{Vir}}$-modules are very interesting tools for explicit computations (these characters are known in statistical physics as "partition functions").

If M is any highest weight $\underline{\text{Vir}}$-module, the associated character is then defined as

$$\text{ch}_M(q) = \sum_{\lambda \in \mathcal{L}'_0} \dim(M_\lambda) q^\lambda$$

It is then fairly easy to see that if M is the Verma module $M(E, z)$, its character is given by

$$\text{ch}_{M(E,z)}(q) = \frac{q^\Lambda}{\varphi(q)},$$

Λ being the highest weight associated to the pair (E, z).

The Verma modules are not the only $\underline{\text{Vir}}$-modules of interest. In the cases where $M(E, z)$ is degenerate, one is interested in its unique irreducible quotient $L(E, z)$; this corresponds to $z = 1$, $E = m^2/4$, and to the discrete series ($z < 1$).

In the case $z = 1$, $E = m^2/4$, the Sugawara construction (see the next section) allows to deduce the character from the corresponding Kac-Moody character [56, 55]:

$$\mathrm{ch}_{L(m^2/4,\ 1)}(q) = \frac{1}{\varphi(q)}\,[q^{m^2/4} - q^{(m+2)^2/4}]$$

Finally, in the case of the discrete series, Rocha-Caridi and Wallach [56] obtained the following result, using again a Jantzen filtration technique:

$$\mathrm{ch}_{L(E_{r,s},z)}(q) = \frac{1}{\varphi(q)} \sum_{k\in\mathbf{Z}} [q^{b(k)} - q^{a(k)}]$$

with

$$a(k) = \frac{[2(m+2)(m+3)k + (m+3)r + (m+2)s]^2 - 1}{4(m+2)(m+3)}$$

$$b(k) = \frac{[2(m+2)(m+3)k + (m+3)r - (m+2)s]^2 - 1}{4(m+2)(m+3)}$$

$z \geq 1$: *The Sugawara Construction*

After having classified the possible places of unitarity for irreducible highest weight $\underline{\text{Vir}}$-modules (characterized by the pair (E, z)), the question arises to know whether these modules are actually unitary or not. The so-called *Sugawara construction* [62] allows us to give a positive answer to this question, in the case $z \geq 1$ (the connection to Kac-Moody representation theory seems to be due to Kac [35]).

The starting point is to consider any untwisted affine Kac-Moody algebra (such a construction works in the twisted case as well, but the exposition is less clear [23, 61]). Let

$$\mathcal{G} = \mathbf{C}[t, t^{-1}] \otimes \mathcal{G}_0 \oplus \mathbf{C}c \oplus \mathbf{C}d$$

and let $\tilde{h}$ be its dual Coxeter number. Let $\{u_i\}$ and $\{u^j\}$ be dual bases of $\mathcal{G}_0$ (i.e., such that $(u_i, u^j)_0 = \delta_i^j$, $(\cdot, \cdot)_0$ being the Killing form of $\mathcal{G}_0$). Finally, define on $\mathcal{U}(\mathbf{C}[t, t^{-1}] \otimes \mathcal{G}_0)$ the following product, called *normal ordered product*. For any $u, v \in \mathcal{G}_0$, $p, q \in \mathbf{Z}$,

$$:(t^p \otimes u)(t^q \otimes v): = \begin{cases} (t^p \otimes u)(t^q \otimes v) & \text{if } q > p \\ \frac{1}{2}[(t^p \otimes u)(t^q \otimes v) + (t^q \otimes v)(t^p \otimes u)] & \text{if } q = p \\ (t^q \otimes v)(t^p \otimes u) & \text{if } p > q \end{cases}$$

Consider any irreducible highest weight $\mathcal{G}$-module $L(\Lambda)$, of level $m \neq -\tilde{h}$, and introduce the following element of $\mathcal{U}(\mathbf{C}[t, t^{-1}] \otimes \mathcal{G}_0)$ acting on $L(\Lambda)$, by

$$\mathcal{L}_n = \frac{1}{2(\tilde{h} + m)} \sum_i \sum_p :(t^p \otimes u_i)(t^{n-p} \otimes u^i):$$

One then has

Proposition 8. The $\mathcal{G}$-module $L(\Lambda)$, with level $m \neq -\tilde{h}$, extends naturally to a highest weight representation of the semidirect product $\mathcal{G} \times \underline{\mathrm{Vir}}$, L_n being represented by the element $\mathcal{L}_n$ of $\mathcal{U}(\mathbf{C}[t, t^{-1}] \otimes \mathcal{G}_0)$, and $\tilde{c}$ by

$$z = \frac{m \dim(\mathcal{G}_0)}{m + \tilde{h}} 1$$

PROOF. One easily checks the following commutation relations: if $n, p \in \mathbf{Z}$, $x \in \mathcal{G}_0$,

$$[\mathcal{L}_n, \mathcal{L}_p] = (n - p)\mathcal{L}_{p+n} + \frac{n^3 - n}{12} \frac{m \dim(\mathcal{G}_0)}{m + \tilde{h}} \delta_{n+p,0}$$

$$[\mathcal{L}_n, t^p \otimes x] = -p(t^{n+p} \otimes x)$$

and the proposition follows. □

One then has the following immediate corollary.

Corollary. Taking for $L(\Lambda)$ an integrable $\mathcal{G}$-module leads to a unitarizable highest weight $\underline{\mathrm{Vir}}$-module, with $z \geq 1$. The particular case $\Lambda = \Lambda_0$ (basic representation) leads to $z = 1$ (this is the original construction of Sugawara [62]).

z < 1: The G.K.O. Construction

In the case of the discrete series, we have

$$z = 1 - \frac{6}{(m + 2)(m + 3)} \quad \text{and}$$

$$E_{r,s} = \frac{[(m + 3)r - (m + 2)s]^2 - 1}{4(m + 2)(m + 3)}$$

Goddard, Kent, and Olive [21] (G.K.O.) gave an explicit construction, based on integrable representations of affine algebras. This construction was extended and simplified later [22, 68, 42, 43]; we adopt here the presentation of [42], handling the untwisted case (which is sufficient to yield the desired result).

Consider two integrable representations of the untwisted affine algebra $\mathcal{G}$, on $L(\Lambda)$ and $L(\Lambda')$, and let m and m' be the respective levels of these representations. The idea of G.K.O. construction is to decompose the tensor product $L(\Lambda) \otimes L(\Lambda')$ into irreducible representations of $\mathcal{G}$.

Lemma 8 (G.K.O.). $L(\Lambda) \otimes L(\Lambda')$ carries a natural unitary highest weight representation of $\underline{\mathrm{Vir}}$, such that L_n is represented by

$$\mathcal{L}_n = \frac{-1}{m + m' + \check{h}} \sum_i \sum_{p \in \mathbf{Z}} : (t^p \otimes u_i) \cdot (t^{n-p} \otimes u^i)$$

and $\tilde{c}$ by

$$\dim(\mathcal{G}_0) \left[\frac{m}{m + \check{h}} + \frac{m'}{m' + \check{h}} - \frac{m + m'}{m + m' + \check{h}} \right] 1$$

(recall that $\{u^i\}$ and $\{u_j\}$ are dual basis of $\mathcal{G}$, and that $\check{h}$ is the dual Coxeter number). This representation commutes with the action of $\mathcal{G}$ on $L(\Lambda) \otimes L(\Lambda')$.

PROOF. The lemma follows from the simple verification of the following relations:

$$[\mathcal{L}_n, \mathcal{L}_p] = (n - p)\mathcal{L}_{n+p} + \frac{\dim(\mathcal{G}_0)}{12}(n^3 - n) \left[\frac{m}{m + \check{h}} + \frac{m'}{m' + \check{h}} - \frac{m + m'}{m + m' + \check{h}} \right] \delta_{n+p,0}$$

$$[\mathcal{L}_n, c] = [\mathcal{L}_n, t^p \otimes x] = 0$$

Moreover, $\mathcal{L}_0$ is diagonal and acts by

$$\mathcal{L}_0 = \frac{1}{2} \left[\frac{(\Lambda, \Lambda + 2\rho)}{m + \check{h}} + \frac{(\Lambda', \Lambda' + 2\rho)}{m' + \check{h}} - \frac{\Omega}{m + m' + \check{h}} \right]$$

Ω being the Casimir element of $\mathcal{G}$, i.e., the element of $\mathcal{U}(\mathcal{G})$ defined by

$$\Omega = \Omega_0 + 2d(c + \check{h}) + \sum_{\substack{p \in \mathbf{Z} \\ p \neq 0}} \sum_i :(t^p \otimes u_i)(t^{-p} \otimes u^i):$$

(Ω_0 is the Casimir element of $\mathcal{G}_0$). □

We remark that, because of the "normal ordering," Ω acts on the highest weight vectors v as

$$\Omega \cdot v = [2(c + \tilde{h})d + 2(\rho, \rho) + h_\rho] \cdot v.$$

Restricting to the particular case $\mathcal{G}_0 = s1(2)$, and $m' = 1$, one has immediately

Corollary. The G.K.O. construction on $s1(2)$ with $m' = 1$ yields all the unitary highest weight representations of $\underline{\text{Vir}}$ in the discrete series.

The result can be made much more precise by looking at the decomposition of the character of $L(\Lambda) \otimes L(\Lambda')$, as follows. We focus on the case $\mathcal{G}_0 = s1(2)$, with $\Lambda' = \Lambda_0(m' = 1)$. The M lattice is then the root lattice Q_0 of $s1(2)$. Denote by α the positive root of $s1(2)$. Λ has then the form

$$\Lambda = m\Lambda_0 + \frac{n}{2}\alpha$$

Denote by $\mathcal{U}_k$ the space spanned by the highest weight vectors of highest weight $\Lambda + \Lambda_0 - k\alpha$; we have

$$\mathcal{U}_k = \{v \in L(\Lambda) \otimes L(\Lambda_0): h_\alpha \cdot v = (n - 2k) \cdot v,\ e_\alpha \cdot v = 0\}$$

$\mathcal{U}^k$ carries a highest weight representation of $\underline{\text{Vir}}$ (since $[\mathcal{L}_n, \mathcal{G}'] = 0, \forall n$); in particular, L_0 acts on $\mathcal{U}^k$ as

$$L_0 = -d + \frac{n(n - 2)}{4(m + 2)} - \frac{(n - 2k)(n + 2 - 2k)}{4(m + 3)}$$

The character of this representation is thus given by

$$\text{ch}_{\mathcal{U}_k}(q) = q^{n(n+2)/4(m+2) - (n-2k)(n+2-2k)/4(m+3)}\text{Tr}(q^d)_{\mathcal{U}_k}$$

in which $\text{Tr}(q^d)_{\mathcal{U}_k}$ can be evaluated from $\text{ch}_{L(\Lambda)}\text{ch}_{L(\Lambda_0)}$, using the elliptic theta functions multiplication rule [38], which reads, for $\mathcal{G}_0 = s1(2)$,

$$\Theta_{m\Lambda_0 + \frac{n\alpha}{2}}\Theta_{m'\Lambda_0 + \frac{n'\alpha}{2}} = \sum_{i \in \mathbf{Z}/(m+m')\mathbf{Z}} \mathcal{M}(m, m', n, n', i)\Theta_{(m+m')\Lambda_0 + (n+n'+2mi)\alpha/2}$$

the multiplicity function $\mathcal{M}$ being given by

$$\mathcal{M}(m, m', n, n', i)(q) = \Theta_{mm'(m+m')\Lambda_0 + (m'n - mn' + 2imm')\alpha/2}$$

Let $r_\Lambda = n + 1$, $s_{\Lambda,k} = n + 1 - 2k$ if $k \geq 0$, and $r_\Lambda = m - n + 1$, $s_{\Lambda,k} = m - n + 2 + 2k$ otherwise. The explicit computation of $\mathrm{ch}_{L(\Lambda_0)}\mathrm{ch}_{L(\Lambda)}$ [42] shows that $\mathcal{U}_k$ is actually an irreducible highest weight $\underline{\mathrm{Vir}}$-module, equivalent to

$$\mathcal{U}_k \cong L(E_{r_\Lambda,s_{\Lambda,k}}(z_m), z_m)$$

(this follows from the identification of the characters).

More precisely, one has [42]

Proposition 9. $L(\Lambda) \otimes L(\Lambda_0)$ splits as a direct sum of $(\mathcal{G}' \oplus \underline{\mathrm{Vir}})$-modules, as follows:

$$L(\Lambda) \otimes L(\Lambda_0) = \sum_{\substack{k \in \mathbf{Z} \\ (n-m-1)/2 \leq k \leq n/2}}^{\oplus} L(\Lambda + \Lambda_0 - k\alpha) \otimes L(E_{r_\Lambda,s_{\Lambda,k}}(z_m), z_m)$$

Some Additional Remarks and Perspectives

(i) The original G.K.O. construction, as proposed by Goddard, Kent, and Olive [21], was slightly more general, as we shall now explain. Let $\mathcal{G}_0$ be a semisimple Lie algebra, and let $\mathcal{G}_0$ be a semisimple $\mathcal{G}_0$-subalgebra. Let $\mathcal{G}$ and $\mathcal{G}'$ be the associated affine Lie algebras (untwisted), and let π be a unitarisable highest weight $\mathcal{G}$-module. Let $\{l_n\}$, $\{l'_n\}$ be the sets of Virasoro generators associated (by the Sugawara construction) to $\pi(\mathcal{G})$ and $\pi(\mathcal{G}')$, and set $l''_n = l_n - l'_n$, $n \in \mathbf{Z}$. The crucial observation is that the l''_n-operators act as a unitarizable highest weight representation of $\underline{\mathrm{Vir}}$, with z-value z'', the difference of the z-values corresponding to $\{l_n\}$ and $\{l'_n\}$. In particular, taking for instance $\mathcal{G}_0 = sp(n)$ and $\mathcal{G}'_0 = sp(n-1) \oplus sp(1)$ one recovers the discrete series of unitary highest weight $\underline{\mathrm{Vir}}$-modules.

A $\mathcal{G}_0$-subalgebra such that $z'' < 1$ (resp. $z'' = 0$) is called a critical (resp. conformal) subalgebra. Conformal and critical subalgebras have been classified by Arcuri [2].

(ii) Conformal field theories are often concerned with supersymmetric extensions of the above results. Essentially, this consists in considering the superaffine algebra [41] $\mathcal{G}^s_\eta$ of any semisimple Lie algebra $\mathcal{G}_0$, defined by

$$\mathcal{G}^s_\eta = \mathbf{C}[t, t^{-1}; \theta] \otimes \mathcal{G}_0 \oplus \mathbf{C}c \oplus \mathbf{C}d$$

where θ is a Grassmann variable ($\theta^2 = 0$) and $\eta = 0$ ("Ramond case") or

$\frac{1}{2}$ ("Neveu-Schwarz case"). The Lie superalgebra structure is given by

$$[t^{n-\eta}\theta \otimes x, t^{m-\eta}\theta \otimes y]_+ = \delta_{m+n,0}(x, y)_0 c, \quad x, y \in \mathcal{G}_0, \quad m, n \in \mathbf{Z}$$

$$[t^n \otimes x, t^{m-\eta}\theta \otimes y] = t^{m+n-\eta}\theta \otimes [x, y]_0, \quad x, y \in \mathcal{G}_0, \quad m, n \in \mathbf{Z}$$

$$[d, t^{n-\eta}\theta \otimes x] = (n - \eta)t^{n-\eta-1}\theta \otimes x, \quad x \in \mathcal{G}_0, \quad n \in \mathbf{Z}$$

$$[c, t^{n-\eta}\theta \otimes x] = 0, \qquad x \in \mathcal{G}_0, n \in \mathbf{Z}$$

together with the usual relations for the affine component of $\mathcal{G}^s_\eta$.

One is also interested in the super Virasoro (or superconformal) algebra

$$\underline{\mathrm{SVir}}^\eta = \underline{\mathrm{Vir}} \oplus \sum_{m \in \mathbf{Z}} \mathbf{C} G_{m-\eta}$$

with relations:

$$[G_{m-\eta}, L_n] = \left(\frac{n}{2} - m + \eta\right) G_{m+n-\eta}, \qquad m, n \in \mathbf{Z}$$

$$[G_{m-\eta}, G_{n-\eta}]_+ = 2L_{m+n} + \frac{\tilde{c}}{3}\left(m^2 - \frac{1}{4}\right)\delta_{m+n,0}, \qquad m, n \in \mathbf{Z}$$

$$[G_{m-\eta}, \tilde{c}] = 0, \qquad m \in \mathbf{Z}$$

The superaffine algebras and super Virasoro algebra have unitary highest weight representation theories quite analogous to the nonsupersymmetric case. In particular, $\underline{\mathrm{SVir}}$ possesss a discrete series of irreducible unitary highest weight representations, parametrized by

$$z = \frac{3}{2}\left(1 - \frac{8}{(m + 2)(m + 4)}\right)$$

$$E_{r,s}(z) = \frac{[(m + 4)r - (m + 2)s]^2 - 4}{8(m + 2)(m + 4)} + \frac{1}{8}\left(\frac{1}{2} - \eta\right)$$

Actually, it turns out that it is possible to do "super Sugawara" and "super G.K.O." constructions [21, 42], which yield results analogous to the nonsupersymmetric case, in particular the unitarity of the representations of the discrete series.

(iii) The G.K.O. construction allows to build infinitely many discrete series of unitarizable highest weight Vir-modules, which are not irreducible. For instance, with $\mathcal{G}_0 = s1(2)$ and $m' = 2$, one gets

$$z = \frac{3}{2}\left[1 - \frac{8}{(m + 2)(m + 4)}\right]$$

or with $\mathcal{G}_0 = s1(3)$ and $m' = 1$,

$$z = 2\left[1 - \frac{12}{(m + 3)(m + 4)}\right]$$

It is therefore reasonable to ask whether such discrete series can correspond to discrete series of irreducible unitarizable highest weight representations of some infinite-dimensional Lie algebra, extension of the Virasoro algebra. Indeed, the answer is positive in some cases: for instance, the discrete series corresponding to $\mathcal{G}_0 = s1(2)$, $m' = 2$, is actually the discrete series of irreducible unitarizable highest weight SVir-modules.

Even in more complicated cases, it is possible to find an extension of Vir realizing this program. For instance, in the case $\mathcal{G}_0 = s1(3)$, $m' = 1$, the discrete series obtained by G.K.O. construction correspond to irreducible unitarizable highest weight representations of the so-called Zamolodchikov algebra [73] (or composite algebra [63]), denoted by Zam.

Zam can be realized as an operator algebra, acting on a space carrying a highest weight representation of Vir, and generated by Vir and additional operators $\{V_n, n \in \mathbf{Z}\}$, with relations

$$[L_n, V_m] = (2n - m)V_{m+n}$$

$$\begin{aligned}[V_n, V_m] = {} & \frac{16}{22 + 5z}(n - m)V_{m+n} \\ & -(m - n)\left(\tfrac{1}{15}(m + n + 2)(m + n + 3) \right. \\ & \left. - \tfrac{1}{6}(m + 2)(n + 2)\right)L_{m+n} \\ & + \frac{z}{360}(m^3 - m)(m^2 - 4)\delta_{m+n,0}\end{aligned}$$

where the Λ_n, $n \in \mathbf{Z}$ are new elements of Zam, defined according to

$$\Lambda_m = \sum_{n\in\mathbf{Z}} : L_n L_{m-n}: + \frac{1}{20}[-m^2 + 5(m \bmod 2) + 4]L_m$$

(: · : is the normal-ordered product previously defined, and m mod 2 stands for the remainder of m modulo 2).

Using tedious but easy computations, one obtains some other relations of Zam, for instance,

$$[L_n, \Lambda_m] = (3n - m)\Lambda_{m+n} - \frac{22 + 5z}{30}(n^3 - n)L_{m+n}$$

and so on.

It is worth noticing that one then obtains a concrete realization of Zam, i.e., a highest weight representation. Actually, Fateev and Zamolodchikov [11] proved that the Verma modules corresponding to the discrete series are unitarizable and degenerate.

Finally, let us quote the important work of Bais et al. [3,4], who used "third order" generalizations of Sugawara [3] and G.K.O. [4] constructions (which are of "second order") for constructing an explicit realization of the V_n generators of Zam. It is now recognized that Zam is a particular case of the so-called extended algebras, appearing in two-dimensional conformal field theories, and realized in the same way as operator algebras. However, it must be said that up to now, very little is known on the representation theory of such algebras. The Zamolodchikov algebra and its various generalizations are also known under the name of W-algebras.

5.3.2 Vertex Operators

Finding explicit realizations of highest weight representations of affine Lie algebras and especially integrable representations, is a very appealing problem. It turns out that such a problem can be solved in some particular cases, namely the case of the basic representation ($\Lambda = \Lambda_0$) of affine Lie algebras built over simply laced $\mathcal{G}_0$'s (i.e. the roots of $\mathcal{G}_0$ have all the same length). The basic tool for this construction is the vertex operator, originally discovered by Skyrme [60]. Vertex operators are actually of great interest in the context of string and superstring theories, where they are the building blocks for the interaction picture (see, e.g., [25] and [1] for an introduction to string theories). The vertex operators were introduced in Kac-Moody representation theory by Lepowsky and Wilson [45] (in the so-called principal realization, i.e., as a representation of loop algebras twisted by the Coxeter element [36]). Their use in the context of string theory was inspired by [57,14], and is known as Frenkel-Kac-Segal construction. Further studies have exhibited the importance of vertex operators in many fields, for instance in connection with the classification of finite simple groups (see, e.g., [16]) or the infinite hierarchies of KP, mKP, KdV, and mKdV equations (see e.g. [40]). The presentation we adopt here is much inspired from [39] [44] [61].

The starting elements of the construction are a simply laced semisimple Lie algebra $\mathcal{G}_0$ and an arbitrary isometry ω of order m ($\omega^m = 1$) of the root lattice Q_0 of $\mathcal{G}_0$. Note that ω extends to an automorphism (also denoted by ω) of $\mathcal{G}_0$, of order $m' = m$ or $2m$. If $\mathfrak{h}_0$ is any Cartan subalgebra of $\mathcal{G}_0$, one has the decomposition:

$$\mathcal{G}_0 = \sum_{p=0}^{m'-1}{}^{\oplus} \mathcal{G}_{0;p}$$

$$\mathfrak{h}_0 = \sum_{p=0}^{m-1}{}^{\oplus} \mathfrak{h}_{0;p}; \quad \mathfrak{h}_{0;p} = \left\{x \in \mathfrak{h}_0 \colon \omega \cdot x = \varepsilon^p x, \ \varepsilon = \exp\left(-\frac{2i\pi}{m}\right)\right\}$$

and one constructs, as usual, the ω-twisted loop algebra:

$$\mathcal{G}_0 = \sum_{p=0}^{m'-1}{}^{\oplus} \mathbf{C}t^p \otimes \mathcal{G}_{0;(p \bmod m')}$$

and its infinite-dimensional Heisenberg subalgebra

$$\mathcal{S} = \sum_{\substack{p\in\mathbf{Z} \\ p\neq 0}}{}^{\oplus} \mathbf{C}t^p \otimes \mathfrak{h}_{0;(p \bmod m)}$$

which admits a natural triangular decomposition $\mathcal{S} = \mathcal{S}_- \oplus \mathcal{S}_0 \oplus \mathcal{S}_+$

The first step is then the construction of a highest weight representation of the Heisenberg subalgebra $\mathcal{S}$. By Stone–von Neumann theorem any such representation is equivalent to a Fock (or "creation-annihilation") representation, built as follows: let $S(\mathcal{S}_-)$ be the symmetric algebra of $\mathcal{S}_-$, and let v be any element of $S(\mathcal{S}_-)$. Then $t^n \otimes h \in \mathcal{S}$ acts on v as follows:

$$(t^n \otimes h) \cdot v = (t^n \otimes h)v \qquad \text{if } n < 0 \quad (\text{left multiplication in } S(\mathcal{S}_-))$$

$$(t^n \otimes h) \cdot v = \partial_{t^{-n}\otimes h} v \qquad \text{if } n > 0$$

Denote by $p_{\omega;k}$ the orthogonal projector on $\mathcal{G}_{0;k}$, and associate with any $\alpha \in Q_0$ the following set of elements of $\mathcal{S}$:

$$\alpha(k) = t^k \otimes p_{\omega;(k \bmod m)}(h_\alpha), \qquad k \in \mathbf{Z}$$

where h_α is the canonical image of α in $\mathfrak{h}_0$, normalized so that $(h_\alpha, h_\alpha) = 2$. It is convenient to introduce the following function $\mathcal{V}$ of the complex variable z, acting on $S(\mathcal{S}_-)$: if $\alpha \in Q_0$

$$\mathcal{V}(\alpha, z) = \exp\left\{-m^{-1/2} \sum_{k<0} \frac{\alpha(k)}{k} z^{-k}\right\} \exp\left\{-m^{-1/2} \sum_{k>0} \frac{\alpha(k)}{k} z^{-k}\right\}$$

This expression is clearly well-defined, and one has, from direct computations, the following result.

Lemma 9.

$$\mathcal{V}(\alpha, z)\mathcal{V}(\beta, z') \underset{|z|>|z'|}{=} :\mathcal{V}(\alpha, z) \cdot \mathcal{V}(\beta, z'): z^{-m(p_{\omega;0}(h_\alpha), p_{\omega:0}(h_\beta))} \cdot \prod_{k=0}^{m-1} (z - \varepsilon^k z')^{(h_\alpha, \omega^k \cdot h_\beta)}$$

The sign $=_{|z|>|z'|}$ means that $\mathcal{V}(\alpha, z)\,\mathcal{V}(\beta, z')$ takes the form of an infinite power series, which converges to the right-hand sign for $|z| > |z'|$.

The same computation leads to

$$\mathcal{V}(\beta, z')\,\mathcal{V}(\alpha, z) \underset{|z'|>|z|}{=} :\mathcal{V}(\alpha, z) \cdot \mathcal{V}(\beta, z'): S(\alpha, \beta) z'^{-m(p_{\omega;0}(h_\alpha), p_{\omega:0}(h_\beta))}$$

the "symmetry factor" $S(\alpha, \beta)$ being defined by

$$S(\alpha, \beta) = \prod_{k=0}^{m-1} (-\varepsilon^k)^{(\omega^k \cdot h_\alpha, h_\beta)}$$

The second step is the introduction of the "zero-mode representation space" $\mathcal{H}$, which is roughly speaking a space carrying a projective action of the root lattice Q_0 of $\mathcal{G}_0$. Introduce first the central extension:

$$1 \to \langle(-1)^m \varepsilon\rangle \to \hat{Q}_0 \xrightarrow{\pi} Q_0 \to 1$$

of Q_0 by the cyclic group $\langle(-1)^m\varepsilon\rangle$ generated by $(-1)^m\varepsilon$, defined by

$$\hat{\alpha}\hat{\beta}\hat{\alpha}^{-1}\hat{\beta}^{-1} = S(\alpha, \beta), \qquad \hat{\alpha}, \hat{\beta} \in \hat{Q}_0$$

$$\pi(\hat{\alpha}) = \alpha, \qquad \pi(\hat{\beta}) = \beta$$

Let $\sigma: Q_0 \to \hat{Q}_0$ be a section of this extension, and denote by $\varepsilon_\sigma: Q_0 \times Q_0 \to \langle(-1)^m\varepsilon\rangle$ the associated 2-cocycle (i.e., ε_σ verifies $\varepsilon_\sigma(\alpha, \beta)\varepsilon_\sigma(\beta, \alpha)^{-1} = S(\alpha, \beta)$ and $\varepsilon_\sigma(\alpha, \beta)\varepsilon_\sigma(\alpha + \beta, \gamma) = \varepsilon_\sigma(\alpha, \beta + \lambda)\varepsilon_\sigma(\beta, \gamma)$ for any $\alpha, \beta, \gamma \in Q_0$), defined by

$$\sigma(\alpha)\sigma(\beta) = \varepsilon_\sigma(\alpha,\beta)\sigma(\alpha + \beta), \qquad \alpha, \beta \in Q_0$$

It is not very difficult to see [44] that there exists an automorphism $\hat{\omega}$ of $\hat{Q}_0$ covering ω, such that

$$\hat{\omega} \cdot \hat{\alpha} = \hat{\alpha} \qquad \text{if } \omega \cdot \pi(\hat{\alpha}) = \pi(\hat{\alpha}), \quad \hat{\alpha} \in \hat{Q}_0$$

$$\hat{\omega} \cdot [(-1)^m\varepsilon] = (-1)^m\varepsilon$$

and for any $\alpha \in Q_0$ and $n \in \mathbf{N}$, there exists $\eta(\alpha, n) \in \langle(-1)^m\varepsilon\rangle$ such that

$$\hat{\omega}^n[\sigma(\alpha)] = \eta(\alpha, n)\sigma[\omega^n \cdot \alpha].$$

Introduce the following Q_0-sublattices:

$$L = (1 - p_{\omega;0})(\mathfrak{h}_0) \cap Q_0, \qquad M = (1 - \omega)Q_0$$

and let $R \supset M$ be the radical of S in L: $R = \{\alpha \in L, S(\alpha, L) = 1\}$. Denote by $\hat{M} \subset \hat{R} \subset \hat{L}$ the associated $\hat{Q}_0$-subgroups. The next lemma then follows [44].

Lemma 10. There exists a unique homomorphism χ: $\hat{M} \to \mathbf{C}$ such that χ fixes $(-1)^m\varepsilon$ and

$$\chi(\hat{\alpha}(\hat{\omega} \cdot \hat{\alpha}^{-1})) = \varepsilon^{-\Sigma_{p=0}^{m-1}(\omega^p \cdot h_\alpha, h_\alpha)/2}, \qquad \hat{\alpha} \in \hat{Q}_0, \quad \pi(\hat{\alpha}) = \alpha$$

Let $\mathscr{E}$ be any irreducible $\hat{L}$-module on which $\hat{M}$ acts by the character χ (it turns out that such $\hat{L}$-modules can be classified [39,44], and that they have dimension $\dim(\mathscr{E}) = \mathrm{Card}(N/R)^{1/2}$), and let $\mathscr{H}$ be the corresponding induced $\hat{Q}_0$-module

$$\mathscr{H} = e^{Q_0} \otimes_{e^L} \mathscr{E}$$

where e^{Q_0} (resp. e^L) denotes the group algebra of Q_0 (resp. L), i.e., the associative algebra generated by the elements e^α, $\alpha \in Q_0$ (resp. $\alpha \in L$), with multiplication law

$$e^\alpha e^\beta = e^{\alpha+\beta}, \qquad \alpha, \beta \in Q_0 \text{ (resp. } L)$$

One needs to define on $\mathscr{H}$ the two following actions: the first one comes from the canonical left action of Q_0 on e^{Q_0}: if $\alpha \in Q_0$

$$\sigma(\alpha) \cdot e^\beta \otimes \xi = \varepsilon_\sigma(\alpha, \beta)e^{\alpha+\beta} \otimes \xi \qquad \text{for any } e^\beta \otimes \xi \in \mathscr{H}$$

The second one comes from the dual action of $p_{\omega;0}(\mathfrak{h}_0)$ on e^{Q_0}: for any $h \in p_{\omega;0}(\mathfrak{h}_0)$

$$\partial_h \cdot e^\beta \otimes \xi = (h, h_\beta)e^\beta \otimes \xi \qquad \forall\, e^\beta \otimes \xi \in \mathscr{H}$$

and one has the following commutation rule:

$$[\partial_h, \sigma(\alpha)] = (h, h_\alpha)\sigma(\alpha), \qquad h \in p_{\omega;0}(\mathfrak{h}), \quad \alpha \in Q_0$$

One is then well prepared to introduce the Vertex operator, acting on $\mathscr{H} \otimes S(\mathscr{S}_-)$. If $\alpha \in Q_0$,

$$\mathscr{U}(\alpha, z) = z^{m|p_{\omega;0}(h_\alpha)|^2/2}\sigma(\alpha)z^{m\partial_{p_{\omega;0}(h_\alpha)}} \otimes \mathscr{V}(\alpha, z)$$

It is fairly easy to see that the action of $\mathcal{U}(\alpha, z)$ on $\mathcal{H} \otimes S(\mathcal{S}_-)$ is well defined, and one can introduce its Laurent coefficients

$$E^i_\alpha = \frac{1}{2i\pi} \oint_0 \frac{dz}{z} z^i\, \mathcal{U}(\alpha, z), \qquad i \in \mathbf{Z}$$

the closed integration contour positively encircling $z = 0$.

Proposition 10. The Laurent coefficients E^i_α satisfy the following commutation relations:

$$[E^i_\alpha, E^j_\beta] = \sum_{\substack{r=0 \\ (h_\alpha, \omega^r h_\beta) = -1}}^{m-1} \eta(\beta, r)\varepsilon^{-rj}\varepsilon'_\sigma(\alpha, \omega^r \cdot \beta) E^{i+j}_{\alpha+\omega^r\cdot\beta}$$

$$- \sum_{\substack{r=0 \\ \omega^r(\beta) = -\alpha}}^{m-1} \eta(\beta, r)\varepsilon^{-rj}\varepsilon'_\sigma(\alpha, -\alpha)[i\delta_{i+j,o} + m\alpha(i+j)],$$

$$\alpha, \beta \in \mathcal{R}_0, \quad i,j \in \mathbf{Z}$$

with

$$\varepsilon'_\sigma(\alpha, \beta) = \varepsilon_\sigma(\alpha, \beta)B(\alpha, \beta)$$

$$B(\alpha, \beta) = \prod_{k=1}^{m-1} (1 - \varepsilon^k)^{(h_\alpha, \omega^k \cdot h_\beta)}$$

PROOF. The proposition follows from an explicit computation of the commutator [61]. As a consequence of the Campbell-Baker-Hausdorff formula, one has

$$[E^i_\alpha, E^j_\beta] = \left(\frac{1}{2i\pi}\right)^2 \left[\oint_0\oint_0{}_{|z|>|u|} - \oint_0\oint_0{}_{|u|>|z|}\right] z^i u^j \frac{dz}{z}\frac{du}{u}\, \mathcal{J}$$

$\mathcal{J}$ being given by

$$\mathcal{J} = z^{m|p_{\omega;0}(h_\alpha)|^2/2} u^{m|p_{\omega;0}(h_\beta)|^2/2} \prod_{k=1}^{m} (z - \varepsilon^k u)^{(h_\alpha, \omega^k \cdot h_\beta)}$$

$$\cdot\, \sigma(\alpha)\sigma(\beta) z^{m\partial p_{\omega;0}(h_\alpha)} u^{m\partial p_{\omega;0}(h_\beta)} \otimes :\mathcal{V}(\alpha, z)\mathcal{V}(\beta, u):$$

By standard arguments, the integration contour can be continuously deformed so that

$$[E^i_\alpha, E^j_\beta] = -\left(\frac{1}{2i\pi}\right)^2 \sum_{r=1}^{m} \oint_0 \frac{dz}{z} \oint_{\varepsilon^{-r}z} \frac{du}{u} z^i u^j \mathcal{J}$$

where the u integration is performed on a contour positively encircling $u = \varepsilon^{-r}z$, excluding $u = 0$. To perform this integration, one has to take into account the contribution of simple poles, for which $(h_\alpha, \omega^r \cdot h_\beta) = -1$, and double poles, for which $(h_\alpha, \omega^r \cdot h_\beta) = -2$, i.e., $\omega^r \cdot h_\beta = -h_\alpha$ (since $\mathcal{G}_0$ is simply laced).

Taking into account the special property of the χ character of $\hat{M}$ (see Lemma 10), one is then led to

$$\begin{aligned}
[E^i_\alpha, E^j_\beta] = & \sum_{\substack{r=1 \\ (h_\alpha,\omega^r h_\beta)=-1}}^{m} \varepsilon^{-rj}\eta(\beta,r)\, \varepsilon_\sigma(\alpha,\omega^r\cdot\beta) \\
& \cdot \prod_{k=1}^{m-1} (1-\varepsilon^k)^{(h_\alpha,\omega^{k+r}h_\beta)}\, E^{i+j}_{\alpha+\omega^r\beta} \\
& + \sum_{r=1}^{m} \varepsilon^{-rj}\eta(\beta,r)\, \varepsilon_\sigma(\alpha,-\alpha) \\
& \cdot \prod_{k=1}^{m-1} (1-\varepsilon^k)^{-(h_\alpha,\omega^k h_\alpha)} [i\delta_{i+j,0} + m\alpha(i+j)]
\end{aligned}$$

and the result follows. □

Comparing such relations with the commutation relations of the ω-twisted loop algebra $\mathcal{G}$, one then has [39,44]

Theorem 9. The operators $\{\alpha(k), k \in \mathbf{Z}, \alpha \in Q_0\}$ (creation-annihilation operators), $\{E^i_\alpha, i \in \mathbf{Z}, \alpha \in Q_0\}$, together with the identity operator, span a Lie algebra of operators acting on the Hilbert space $\mathcal{H} \otimes S(\mathcal{S}_-)$, is isomorphic with the ω-twisted loop algebra

$$\mathcal{G} = \sum_{n\in\mathbf{Z}}^{\oplus} \mathbf{C}t^n \otimes \mathcal{G}_{0;n \bmod m'} \oplus \mathbf{C}c \oplus \mathbf{C}d$$

Moreover, their action on $\mathcal{H} \otimes S(\mathcal{S}_-)$ is irreducible (as a consequence of the irreducibility of the $\hat{L}$-module ξ), and is equivalent to the basic representation of $\mathcal{G}$.

REMARK. For a fixed $\mathcal{G}_0$, if ω runs over the Weyl group W_0 of $\mathcal{G}_0$, one then obtains card(W_0) equivalent representations of the corresponding $\mathcal{G}$. However, finding the corresponding intertwinners seems to be a difficult problem, not yet solved, which can also have important applications in the context of string theories.

5.4 BEYOND AFFINE LIE ALBEGRAS

Affine Lie algebras have been shown to possess a very rich (highest weight) representation theory, and as a consequence, a very large domain of applications in mathematics and physics. Let us quote for instance the domain of two-dimensional conformal field theories and string theories, in which the techniques described in the previous sections play a central role. It is thus relevant to ask whether it is possible to generalize some of these constructions to a larger class of infinite-dimensional Lie algebras in order to investigate for example higher-dimensional quantum field theories, membrane theories or many other potential fields of application. Frenkel recently reviewed some of the existing ways of generalizing affine Lie algebras [13] (the title of the current section is borrowed from him); one can study for instance the Kac-Moody algebras whose G.C.M. has a kernel of more than one dimension, but very little is known about them; it seems that actually the existence of a G.C.M. is a much too strong assumption. Generalization via geometry has been studied in particular by Mickelsson [49] and Singer [59], who exhibited noncentral extensions of current groups and algebras. We refer to [54] for this geometrical approach. We also mention generalizations of the "quantum group" type [10]. We are essentially interested here in generalizations of the type "current algebras," for some of which it is possible to constrct a structure theory (quasisimple Lie algebras) [66,27], and investigate the unitarizable highest weight representation theory [67].

5.4.1 Quasisimple Lie Algebras: Structure Theory

This subsection is devoted to the description of the structure theory of quasisimple Lie algebras and to the classification of their root systems. Most of the proofs involved are fairly standard and will not be reproduced here. The interested reader will find them in [27] and in [6,5,17,28,69] for the semisimple case.

Definition 5. A *quasisimple Lie algebra* is a complex Lie algebra $\mathcal{G}$ such that:

(i) $\mathcal{G}$ is provided with a nondegenerate invariant symmeric bilinear form, called the *Killing form*, and denoted by $(.\ .\ .)$. For simplicity, the Killing form will be assumed to be real on the real linear span of the root system.

(ii) $\mathcal{G}$ possesses a Cartan subalgebra $\mathfrak{h}$, such that

(i) $\mathfrak{h}$ is abelian, and its adjoint action on $\mathcal{G}$ is diagonalizable,
(ii) $\mathrm{Ad}(\mathfrak{h})$ has discrete spectrum,
(iii) with respect to $\mathrm{Ad}(\mathfrak{h})$, $\mathcal{G}$ admits a rootspace decomposition:

$$\mathcal{G} = \mathfrak{h} \oplus \sum_{\alpha \in \mathfrak{R}}^{\oplus} \mathcal{G}_\alpha.$$

where $\mathcal{R} = \mathrm{Sp}(\mathrm{Ad}(\mathfrak{h}))$ is the root system and $\mathcal{G}_\alpha = \{g \in \mathcal{G} | [h, g] = \alpha(h)g \ \forall\, h \in \mathfrak{h}\}$ is the rootspace attached to α. (Note that $\mathfrak{h} = \mathcal{G}_0$.)

Let us postpone the end of Definition 5, for which we need the following remarks and definition.

REMARKS. The following properties are easily proven.

(i) For any $\alpha, \beta \in \mathcal{R}$, the invariance of the Killing form implies that $(\mathcal{G}_\alpha, \mathcal{G}_\beta) = 0$ if $\beta \neq -\alpha$.

(ii) The root system is symmetric: $-\mathcal{R} = \mathcal{R}$

(iii) $\forall\, \alpha, \beta \in \mathcal{R}$: $[\mathcal{G}_\alpha, \mathcal{G}_\beta] \subseteq \mathcal{G}_{\alpha+\beta} = \{0\}$ if $\alpha + \beta \notin \mathcal{R}$).

(iv) For any $\lambda \in \mathfrak{h}'$, let $h_\lambda \in h$ be defined by $(h_\lambda, h) = \lambda(h) \ \forall\, h \in \mathfrak{h}$.

The Killing form can thus be carried onto $\mathfrak{h}'$ as follows: for any $\lambda, \mu \in \mathfrak{h}'$, let us define $(\lambda, \mu) = (h_\lambda, h_\mu)$. $(. \ . \ .)$ is nondegenerate on $\mathfrak{h}'$, and will be called the *Killing form* too.

Definition 6. A root $\delta \in \mathcal{G}$ is called *isotropic* (or *null*) if $(\delta, \delta) = 0$.

Definition 5 (continued)

(iii) For any nonisotropic $\alpha \in \mathcal{R}$, $\mathrm{Ad}(\mathcal{G}_\alpha)$ is nilpotent.

REMARK. Assumption (iii) is not fulfilled by indefinite Kac-Moody algebras (see Theorem 1).

Let us fix notation. From now on, $\mathcal{G}$ is a quasisimple Lie algebra, with Cartan subalgebra $\mathfrak{h}$, Killing form $(. \ . \ .)$ and root system $\mathcal{R}$. It is not difficult to see that the behavior of the nonisotropic roots is the same as the behavior of the roots of semisimple Lie algebras. In particular, one is led to the following result.

Proposition 11. Let be any nonisotropic root in $\mathcal{R}$. One then has

(i) $\dim(\mathcal{G}_\alpha) = 1$.

(ii) For any $\beta \in \mathcal{R}$: $2(\alpha, \beta)/(\alpha, \alpha) \in \mathbf{Z}$.

(iii) Defining the Weyl reflection w_α associated with α by

$$w_\alpha \cdot \lambda = \lambda - 2\frac{(\alpha, \lambda)}{(\alpha, \alpha)}\alpha \qquad \forall\, \lambda \in \mathfrak{h}'$$

then $w_\alpha \cdot \mathcal{R} = \mathcal{R}$.

(iv) $k\alpha \in \mathcal{R}$ if and only if $k = 0$ or ± 1.

(v) For any $\beta \in \mathcal{R}$, there exists a pair (n_-, n_+) of nonnegative integers, such that $\beta + n\alpha \in \mathcal{R}$ if and only if $-n_- \leq n \leq n_+$, n_- and n_+ being related by

$$n_- - n_+ = 2\frac{(\alpha, \beta)}{(\alpha, \alpha)}$$

$\{\beta + n\alpha, n = -n_-, 1 - n_-, \ldots, n_+\}$ is called the *α-string of roots* through β.

PROOF. The proposition basically follows from the observation that if α is nonisotropic, then $\mathcal{G}_\alpha$, $\mathcal{G}_{-\alpha}$ and $[\mathcal{G}_\alpha, \mathcal{G}_{-\alpha}]$ span a complex Lie algebra isomorphic to sl(2,**C**); the proposition is thus a consequence of sl(2,**C**) representation theory. □

Kac [32,31], Moody [50] and McDonald [48] pointed out that the infinite-dimensional structure of affine Lie algebras lies, as far as the root system is concerned, in the existence of isotropic roots. This fact is illustrated by the following discussion.

Lemma 11. Let δ be any isotropic root in $\mathcal{R}$. Then $(\delta, \alpha) = 0\ \forall\ \alpha \in \mathcal{R}$.

Corollary. If $\mathcal{G}$ possesses isotropic roots, the nondegeneracy of the Killing form implies that the dual $\mathfrak{h}'$ of $\mathfrak{h}$ contains more than the complex linear span of $\mathcal{R}$.

PROOF. The proof of Lemma 11 uses the following result.

Lemma 12. Let δ be an isotropic root of $\mathcal{G}$. If $(\delta, \alpha) \neq 0$ for some nonisotropic $\alpha \in \mathcal{R}$, then $\alpha + n\delta \in \mathcal{R}$ for infinitely many consecutive integral numbers n.

PROOF. Let $p \in \mathbf{Z}$ such that $\alpha' = \alpha + p\delta \in \mathcal{R}$, and $\alpha' - \delta \notin \mathcal{R}$; let x_α and x_δ be nonzero elements in $\mathcal{G}_\alpha$ and $\mathcal{G}_\delta$ respectively. Set $x_0 = x_\alpha$, and $x_n = (\mathrm{Ad}(x_\delta))^n \cdot x_0$. One then has, for $x_{-\delta} \in \mathcal{G}_{-\delta}$ normalized so that $(x_{-\delta}, x_\delta) = 1$,

$$(x_{-\delta}, x_1) = -(\delta, \alpha)x_0$$

By the induction procedure, one shows that

$$(x_{-\delta}, x_1) = -n(\delta, \alpha)x_{n-1}$$

and then that $x_n \neq 0$ if $x_{n-1} \neq 0$, and Lemma 12 follows. □

It is now possible to prove Lemma 11. Assuming $(\delta, \alpha) \neq 0$, set $\gamma_n = w_{\alpha'+n\delta}\delta$. Then

$$\gamma_n = \delta - 2\frac{(\delta, \alpha)}{(\alpha', \alpha') + 2n(\delta, \alpha)}(\alpha' + n\delta)$$

and the family $\{\gamma_n\}_{n\in N}$ of elements of $\mathcal{R}$ (by Lemma 12) has a limit point at 0, contradicting hypothesis (ii) of Definition 5. The lemma is thus proved. □

The separation of the isotropic part of $\mathcal{R}$ can be performed as follows: let $\mathfrak{h}'_{\mathcal{R}}$ be the real linear span of $\mathcal{R}$ and $\mathfrak{h}_{\mathcal{R}}$ its real dual space. Introduce the linear mapping $\mathfrak{l}: \mathfrak{h}'_{\mathcal{R}} \to \mathfrak{h}_{\mathcal{R}}$ defined by

$$\mathfrak{l}(\lambda) \cdot (\mu) = (\lambda, \mu) \quad \forall\mu \in \mathfrak{h}'_{\mathcal{R}} \qquad \text{if } \lambda \in \mathfrak{h}'_{\mathcal{R}}$$

and denote for simplicity

$$\mathfrak{l}_\lambda = \mathfrak{l}(\lambda), \qquad \lambda \in \mathfrak{h}'_{\mathcal{R}}$$

$$\mathfrak{h}_* = \mathfrak{l}(\mathfrak{h}'_{\mathcal{R}}), \qquad \mathcal{R}'_{\mathfrak{l}} = \mathfrak{l}(\mathcal{R})$$

$\mathfrak{l}$ and the Killing form restricted to $\mathfrak{h}'_{\mathcal{R}}$ induce on $\mathfrak{h}_*$ a symmetric bilinear form, which has the following easily proven property.

Lemma 13. (. , .) is real and nondegenerate on $\mathfrak{h}_*$.

Clearly from Lemma 11, the $\mathfrak{l}$ mapping sends the isotropic part of $\mathcal{R}$ to zero.

In the finite-dimensional semisimple case, all the roots are nonisotropic, and their complex linear span is noting but the dual $\mathfrak{h}'$ of $\mathfrak{h}$. Moreover, the Killing form is positive-definite on $\mathfrak{h}'_{\mathcal{R}}$. It is actually not difficult to see that the quasisimple Lie algebras whose restriction of the Killing form to $\mathfrak{h}'_{\mathcal{R}}$ is positive-definite exhaust the finite-dimensional semisimple Lie algebras. This follows from the classification of the finite root systems, e.g. [5]. To go further, let us specialize to the nonsemisimple case and introduce the following:

Definition 7. Let $\mathcal{G}$ be a nonsemisimple quasisimple Lie algebra, with the above notations. $\mathcal{G}$ is said to be of *elliptic type* if the Killing form is positive semidefinite on $\mathfrak{h}'_{\mathcal{R}}$ (otherwise, $\mathcal{G}$ is said to be of *indefinite type*). In this case, let $\mathfrak{h}'_*$ be the real predual space of $\mathfrak{h}_*$, and let $\mathfrak{k}_{\mathcal{R}} = \ker(\mathfrak{l})$. Then $\mathfrak{h}'_{\mathcal{R}} = \mathfrak{k}_{\mathcal{R}} \oplus \mathfrak{h}'_*$. $\nu = \dim(\mathfrak{k}_{\mathcal{R}})$ is called the *type* of $\mathcal{G}$, and $n + \nu = \dim(\mathfrak{h}'_{\mathcal{R}})$ the *rank* of $\mathcal{G}$.

REMARK. As we will see, the case $\nu = 1$ corresponds to affine Kac-Moody algebras, and in some cases with $\nu \geq 2$, one can find some current algebras realizations.

It is interesting to mention that up to now, there does not seem to by any known realization of a quasisimple Lie algebra of the indefinite type. Kac mentioned that the indefinite class could actually be empty. This remains an appealing still open problem.

From now on, let us focus on the elliptic case. Introduce the Weyl reflections in $\mathcal{R}'_{\mathfrak{l}}$ by

$$w_{\mathfrak{l}_\alpha} \cdot \mathfrak{l}_\beta = \mathfrak{l}_{w_\alpha} \cdot \beta \qquad \text{for any } \alpha, \beta \in \mathcal{R}, \alpha \text{ nonisotropic}$$

and one easily checks that

$$w_{\mathfrak{l}_\alpha} \cdot \mathfrak{l}_\beta = \mathfrak{l}_\beta - 2\,\frac{(\alpha, \beta)}{(\alpha, \alpha)}\,\mathfrak{l}_\alpha$$

which shows the following.

Proposition 12. $\mathcal{R}'_{\mathfrak{l}}$ is a finite root system, in the following sense [5]:

(i) $\mathcal{R}'_{\mathfrak{l}}$ generates $\mathfrak{h}_*$ and is finite.

(ii) For any $\mathfrak{l}_\alpha, \mathfrak{l}_\beta \in \mathcal{R}'_{\mathfrak{l}}, \mathfrak{l}_\alpha \neq 0$, $2(\mathfrak{l}_\alpha, \mathfrak{l}_\beta)(\mathfrak{l}_\alpha, \mathfrak{l}_\alpha)$ is an integral number.

(iii) $\mathcal{R}'_{\mathfrak{l}}$ is Weyl invariant: $w_{\mathfrak{l}_\alpha} \cdot \mathfrak{l}_\beta \in \mathcal{R}'_{\mathfrak{l}}$ for any $\mathfrak{l}_\alpha, \mathfrak{l}_\beta \in \mathcal{R}'_{\mathfrak{l}}$, $\mathfrak{l}_\alpha \neq 0$. $\mathcal{R}'_{\mathfrak{l}}$ is called the *gradient coroot system*.

REMARK. Contrary to the case of root systems of finite-dimensional semisimple Lie algebras, $\mathcal{R}'_{\mathfrak{l}}$ is not necessarily reduced. In fact, there may exist $\mathfrak{l}_\alpha \in \mathcal{R}'_{\mathfrak{l}}$ such that $2\mathfrak{l}_\alpha \in \mathcal{R}'_{\mathfrak{l}}$. This simply means that if α is a preimage of $\mathfrak{l}_\alpha$ in $\mathcal{R}$, there exists $\eta \in \mathfrak{k}_{\mathcal{R}}$ such that $2\alpha + \eta \in \mathcal{R}$. The following result is useful and will be frequently used below.

Lemma 14. Let $\alpha \in \mathcal{R}$, α nonisotropic, and let $\delta \in \mathfrak{k}_{\mathcal{R}}$. Then if $\alpha + \delta \in \mathcal{R}$, one also has that $\alpha - \delta \in \mathcal{R}$, $\delta - \alpha \in \mathcal{R}$, and $\delta \in \mathcal{R}$.

PROOF. The proof of the lemma follows from $sl(2, \mathbf{C})$ representation theory. $\alpha + \delta + n\delta \in \mathcal{R}$ if and only if $-n_- \leq n \leq n_+$ and $n \in \mathbf{Z}$, with

$$n_- - n_+ = 2\frac{(\alpha + \delta, \alpha)}{(\alpha, \alpha)} = 2.$$

Then, $n_- \geq 2$, and the lemma follows. □

Before giving a more precise description of $\mathcal{R}$, one needs to introduce a notion of irreducibility for quasisimple root systems.

Definition 8. An elliptic quasisimple root system $\mathcal{R}$ is said to be *irreducible* if it has the two following properties:

(i) the gradient coroot system $\mathcal{R}'_{\mathfrak{l}}$ is irreducible

(ii) any isotropic root δ is *unisolated* in the following sense: there exists $\alpha \in \mathcal{R}$, α nonisotropic, such that $\alpha + \delta \in \mathcal{R}$.

In what follows, we specialize to the study of ireducible quasisimple root systems; let us first focus on the cases where the gradient coroot system $\mathcal{R}'_{\mathfrak{l}}$ is a reduced finite root system.

The "reduced case"

Let $\mathfrak{l}_1, \ldots \mathfrak{l}_n$ be a basis of simple roots of $\mathcal{R}'_{\mathfrak{l}}$. It clearly defines on $\mathfrak{h}'_{\mathcal{R}}$ an equivalence relation (for any $\lambda, \mu \in \mathfrak{h}'_{\mathcal{R}}$: $\lambda \sim \mu$ if $\mathfrak{l}_\lambda = \mathfrak{l}_\mu$), $\mathcal{R}'_{\mathfrak{l}}$ being canonically identified to the quotient space. Choose in each equivalence class $\mathfrak{l}_i$ $(i = 1, \ldots, n)$ a representative element α_i such that $\alpha_i \in \mathcal{R}$ $\{\alpha_i, i = 1, \ldots, n\}$ is easily seen to be a basis of simple roots of an irreducible finite root system $\mathcal{R}_{\mathfrak{l}}$, isomorphic to $\mathcal{R}'_{\mathfrak{l}}$ which is a subroot system of $\mathcal{R}$. $\mathcal{R}_{\mathfrak{l}}$ is called the *gradient root system*.

REMARKS. 1. Such a choice for the representative elements α_i is always possible, but far from being unique. Other choices, in which some of the α_i do not belong to $\mathcal{R}$ are in fact also possible. If $\nu = 1$, such choices correspond to different gradations of the affine Lie algebra. Note that the expression "gradient root system" has here a different meaning than in McDonald's terminology [48].

We now turn to the description of $\mathcal{R}$. Fix $\delta \in \mathfrak{t}_{\mathcal{R}}$; let $\alpha \in \mathcal{R}$, nonisotropic, be such that $\alpha + \delta \in \mathcal{R}$, and define ξ_α to be the smallest element on the half-line $\mathbf{R}^+\delta$ such that $\beta = \alpha + \xi_\alpha \in \mathcal{R}$. For any $r \in \mathbf{R}$ and $m \in \mathbf{N}$

$$(w_\beta \cdot w_\alpha)^m(\alpha + r\xi_\alpha) = \alpha + (2m + r)\xi_\alpha.$$

Setting $r = 0$ and 1 in this equality shows that $\alpha + n\xi_\alpha \in \mathcal{R}$ for any $n \in \mathbf{Z}$.

Conversely, m can be chosen in such a way that the minimality assumption made on ξ_α is contradicted if r is not integral. Summarizing, one has

Lemma 15. $\alpha + n\xi_\alpha \in \mathcal{R}$ if and only if n is an integral number.

Consider now the Dynkin diagram of $\mathcal{R}_{\mathfrak{f}}$; by the irreducibility of $\mathcal{R}$, it is connected, and consider an arbitrary connected two-vertices subdiagram, of the form

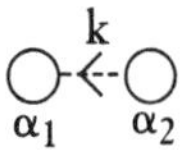

(k = 1, 2, or 3)

which means that $(\alpha_2, \alpha_2) = k(\alpha_1, \alpha_1)$. Introduce as above the elements ξ_{α_1} and ξ_{α_2}. Then one has

$$(w_{\alpha_1} \cdot w_{\alpha_1 + \xi_{\alpha_1}}) \cdot \alpha_2 = \alpha_2 + k\xi_{\alpha_1} \in \mathcal{R}$$

$$(w_{\alpha_2} \cdot w_{\alpha_2 + \xi_{\alpha_2}}) \cdot \alpha_1 = \alpha_1 + \xi_{\alpha_2} \in \mathcal{R}$$

(The existence of ξ_{α_2} follows from the first equality.) One then has $k\xi_{\alpha_2} \in \mathbf{N}\xi_{\alpha_1}$ and $\xi_{\alpha_1} \in \mathbf{N}\xi_{\alpha_2}$, from which one deduces

Lemma 16. Assuming that ξ_{α_1} is known, there are at most two possibilities for ξ_{α_2}:

1. $\xi_{\alpha_2} = \xi_{\alpha_1}$: the direction defined by δ is said to be *untwisted*.
2. $\xi_{\alpha_2} = k\xi_{\alpha_1}$: the direction defined by δ is said to be *twisted*.

This terminology will become clearer when considering the current algebras realizations.

Corollary. (i) If $\mathcal{R}_{\mathfrak{f}}$ is simply laced ($k = 1$ for each connected two-vertices subdiagram), there can be no twist.

(ii) If δ is any isotropic root of $\mathcal{G}$, and α any element of $\mathcal{R}_{\mathfrak{f}}$ of minimal length then $\alpha + \delta \in \mathcal{R}$.

Summarizing, it appears that the most general form of the elements of $\mathcal{R}$ is

$$\gamma_i = \delta_s \qquad \text{(isotropic roots)}$$

$$\gamma_s = \alpha_s + \delta_s \qquad \text{(short roots)}$$

$$\gamma_L = \alpha_L + \delta_L \qquad \text{(long roots)}$$

where α_s (resp. α_L) is a short (resp. long) element in $\mathcal{R}_{\mathfrak{l}}$, and δ_s (resp. δ_L) lies in some lattice Λ_s (resp. Λ_L) contained in $\mathfrak{t}_{\mathcal{R}}$. Moreover, one has $\Lambda_L \subseteq \Lambda_s$.

It is necessary to better describe the Λ_L and Λ_s lattices. Let us start by giving some general results about lattices and their **Z**-basis (see [27]).

Definition 8. Let δ be any element in some lattice Λ. δ is said to be *minimal* if for any natural number p different than 0 or 1, $\delta/p \notin \Lambda$.

Fix a family $\delta_1, \ldots, \delta_\nu$ ($\nu = \dim(\Lambda)$) of **R**-linearly independent elements of Λ, and associate with any family $\{m\} = \{m_1, \ldots, m_\nu\}$ of integral numbers the element

$$\delta_{\{m\}} = \sum_{i=1}^{\nu} m_i\delta_i$$

Lemma 17. $\{\delta_j\}_{j=1,\ldots,\nu}$ is a **Z**-basis of Λ if and only if for any family $\{m_j\}_{j=1,\ldots,\nu}$ of relatively prime integral numbers, $\delta_{\{m\}}$ is minimal.

PROOF. The proof goes as follows: $\delta_{\{m\}}$ is not minimal if and only if one can find some natural number p different from 0 or 1 such that $\delta_{\{m\}}/p \in \Lambda$. But $\delta_{\{m\}}/p = \Sigma_{i=1}^{\nu} m_i\delta_i/p$, and the lemma follows from the assumption that the $\{m_i\}_{i=1,\ldots,\nu}$ are relatively prime. □

Assume now that $\{\delta_j\}_{j=1,\ldots,\nu}$ is any **Z**-basis of Λ, and let M be a $\nu \times \nu$ matrix with integral coefficients. Define

$$\psi_i = \sum_{i=1}^{\nu} M_{ij}\delta_j, \qquad i = 1, \ldots, \nu$$

Clearly, $\{\chi_i\}_{j=1,\ldots,\nu}$ can be a **Z**-basis for Λ only if Det $(M) = 1$.

Return now to the special cases of the Λ_s and Λ_L lattices. Pick any **Z**-basis $\{\delta_j\}_{j=1,\ldots,v}$ of Λ_s; the δ_i's are roots, and the most general form of isotropic and short roots is

$$\gamma_i = \sum_{j=1}^{\nu} n_i\delta_i$$

$$\gamma_s = \alpha_s + \sum_{j=1}^{\nu} n_i\delta_i$$

where α_s is a short element in $\mathcal{R}_{\mathfrak{l}}$, and the n_i's ($i = 1, \ldots, \nu$) run over **Z**.

Moreover, the most general form of the long roots of $\mathcal{R}$ is

$$\gamma_L = \alpha_L + \sum_{j=1}^{\nu} \left[\sum_{i=1}^{\nu} a_i^{(j)} m_i \right] \delta_j,$$

where α_L is some generic long element of $\mathcal{R}_l$, the m_i's ($i = 1, \ldots, \nu$) run over $\mathbf{Z}$, and the $a_i^{(j)}$ are integral coefficients, to be specified below. Denote by k_i the greatest common denominator of the family $\{a_i^{(j)}\}_{j=1,\ldots,\nu}$ and set

$$\xi^i = \frac{1}{k_i} \sum_{j=1}^{\nu} a_i^{(j)} \delta_j.$$

By assumption, $\{\xi^i\}_{i=1,\ldots,\nu}$ is a $\mathbf{Z}$-basis of Λ_s, and the most general form of the elements of $\mathcal{R}$ reads

$$\gamma_i = \sum_{i=1}^{\nu} n_i \xi^i$$

$$\gamma_s = \alpha_s + \sum_{i=1}^{\nu} n_i \xi^i$$

$$\gamma_L = \alpha_L + \sum_{i=1}^{\nu} n_i k_i \xi^i$$

with the same notations as above. Moreover, by Lemma 16, one knowns that $k_i = 1$ or k ($i = 1, \ldots, \nu$). Define the integral number $0 \leq \tau \leq \nu$ by $\Pi_{i=1}^{\nu} k_i = k^{\tau}$. τ is called the *twist number*, and one has:

Lemma 18. Let $\{\psi^i\}_{i=1,\ldots,\nu}$ be another $\mathbf{Z}$-basis of Λ_s, such that $\{l_i \psi^i\}_{i=1,\ldots,\nu}$ is a $\mathbf{Z}$-basis of Λ_L, and let τ' be the associated twist number. Then $\tau' = \tau$.

PROOF. Let P be the $\nu \times \nu$ matrix defined by

$$\psi^i = \sum_{i=1}^{\nu} p_{ij} \xi^j, \qquad i = 1, \ldots \nu$$

Then det $(P) = 1$; moreover, one has that

$$l_i \psi_i = \sum_{j=1}^{\nu} \frac{l_i}{k_j} p_i k_j \xi^j.$$

$$\text{Det} \left\{ \frac{l_i}{k_j} p_{ij} \right\}_{i,j=1,\ldots,\nu} = \frac{\Pi_{j=1}^{\nu} l_i}{\Pi_{j=1}^{\nu} k_j} \text{Det}(P) = 1$$

implies that

$$\frac{\Pi_{j=1}^{\nu} l_i}{\Pi_{j=1}^{\nu} k_j} = k^{\tau' - \tau} = 1$$

and the lemma follows. □

Summarizing, one can state the following.

Theorem 10 [27,66]. Let $\mathcal{R}$ be an irreducible quasisimple root system of the elliptic type, with reduced gradient root system $\mathcal{R}_{\mathfrak{l}}$. $\mathcal{R}$ is then completely characterized by $\mathcal{R}_{\mathfrak{l}}$., its type ν and its twist number τ (ν, $\tau \in \mathbf{N}$, $0 \leq \tau \leq \nu$), from which $\mathcal{R}$ can be described by

$$\mathcal{R} = \{n_i\xi^i,\ n_i \in \mathbf{Z}\ (i = 1, \ldots, \nu)\}$$

$$\dot{\cup}\left\{\alpha_s + \sum_{i=1}^{\nu} n_i\xi^i,\ \alpha_s \in (\mathcal{R}_{\mathfrak{l}})_s,\ n_i \in \mathbf{Z}\ (i = 1, \ldots, \nu)\right\}$$

$$\dot{\cup}\left\{\alpha_L + \sum_{i=1}^{\tau} 2n_i\xi^i + \sum_{i=\tau+1}^{\nu} n_i\xi^i,\ \alpha_L \in (\mathcal{R}_{\mathfrak{l}})_L,\right.$$

$$\left. n_i \in \mathbf{Z}\ (i = 1, \ldots, \nu)\right\}$$

The situation is summarized by the notation $\mathcal{R} \approx (\mathcal{R}_{\mathfrak{l}};\ \nu,\ \tau)$.

The "Nonreduced Case"

Consider now the case where $\mathcal{R}_{\mathfrak{l}}$ is a nonreduced finite root system. From the irreducibility of $\mathcal{R}$, it follows that $\mathcal{R}_{\mathfrak{l}}$ is of the type $\mathbf{BC}_n$, i.e., its Dynkin diagram is of the form

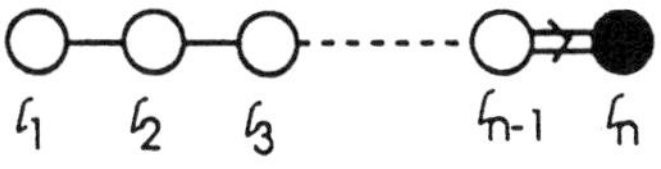

the black vertex • $\mathfrak{l}_n$ meaning that $\mathfrak{l}_n \in \mathcal{R}'_{\mathfrak{l}}$ and $2\,\mathfrak{l}_n \in \mathcal{R}'_{\mathfrak{l}}$ $\mathfrak{l}_n$ being the only simple root having such a property (see [5] for a complete description of nonreduced root systems). As in the previous case, $\mathcal{R}'_{\mathfrak{l}}$ can be seen as a quotient set of $\mathcal{R}$ and one has to choose a representative element in each class. Associate to any $\mathfrak{l}_i$ ($i = 1, \ldots, n$) an $\alpha_j \in \mathcal{R}$ (such a choice is always possible), and associate to $2\mathfrak{l}_n$ the representative $2\alpha_n$, which is obviously not a root in $\mathcal{R}$; it is however possible to find $\eta \in \mathfrak{k}_{\mathcal{R}}$ such that $2\alpha_n + \eta \in \mathcal{R}$. It follows from the elementary properties of the Weyl groups of $\mathcal{R}$ and $\mathcal{R}'_{\mathfrak{l}}$, that

(i) The α_i's, $i = 1, \ldots, n$, form a basis of simple roots of a $\mathbf{B}_n$ root subsystem of $\mathcal{R}$.

(ii) $(\alpha_1, \ldots, \alpha_n, 2\alpha_n\}$ is a basis of simple roots of the gradient root system, $\mathcal{R}'_{\mathfrak{l}} = \mathbf{BC}_n$, which is not included in $\mathcal{R}$.

(iii) For any short root $\alpha_s \in \mathcal{R}'_{\mathfrak{l}}$, $2\alpha_s + \eta \in \mathcal{R}$. One now needs an analogue of Lemma 15; fix any $\delta \in \mathfrak{k}_R$, and let α be any short root in $\mathcal{R}$; define ξ_α to be the smallest element in $\mathbf{R}^+\delta$ such that $2\alpha + \xi'_\alpha \in \mathcal{R}$. One then has

Lemma 19. (i) $2\alpha + r\xi'_\alpha \in \mathcal{R}$ if and only if r is an odd integral number.

(ii) For any short root $\alpha \in \mathcal{R}$, $\xi'_\alpha = \xi_\alpha$

PROOF. $w_\alpha \cdot (2\alpha + \xi'_\alpha) = -2\alpha + \xi'_\alpha \in \mathcal{R}$. Then $2\alpha - \xi'_\alpha \in \mathcal{R}$, and (i) follows from Lemma 15. Moreover, since $w_{2\alpha+\xi'_\alpha} \cdot \alpha = -\alpha + \xi'_\alpha$, it follows that ξ'_α is a nonzero integral multiple of ξ_α: $\xi'_\alpha = p\,\xi_\alpha$, and p is obviously odd. Finally, assuming $p \geq 3$, it is then possible to find a positive integral number $p/4 < m < p/2$. Then $w_{2\alpha+m\xi'_\alpha} \cdot (2\alpha + \xi'_\alpha) = -(2\alpha + (4\,m - p)\,\xi'_\alpha) \in \mathcal{R}$ which is in contradiction with the minimality of ξ'_α; then $p = 1$, which proves the lemma. □

From the same arguments, $w_{\alpha_n} \cdot (2\alpha_n + \eta) = -2\alpha_n + \eta$, one deduces that $\eta \in \mathcal{R}$. Summarizing, it appears that the generic form of the elements of $\mathcal{R}$ is the following:

$$\gamma_i = \delta_s \qquad \text{(isotropic roots)}$$

$$\gamma_s = \alpha_s + \delta_s \qquad \text{(short roots)}$$

$$\gamma_L = \alpha_L + \delta_L \qquad \text{(long roots)}$$

$$\gamma_{sL} = 2\alpha_s + \delta_{sL} \qquad \text{(``superlong'' roots)}$$

where α_s (resp. α_L) is a short (resp. long) root in $\mathcal{R}_{\mathfrak{l}}$, and δ_s (resp. δ_L) (resp. δ_{sL}) lies in some lattice Λ_s (resp. Λ_L) (resp. Λ_{sL}) imbedded in $\mathfrak{k}_{\mathcal{R}}$. Moreover, $\Lambda_{sL} \subseteq \Lambda_L \subseteq \Lambda_s$. It is then possible to carry out the same analysis as in the previous subsection (the "reduced case"), both for the pair (Λ_s, Λ_L) and the pair $(\Lambda_L, \Lambda_{sL})$. One is then provided with a $\mathbf{Z}$-basis $\{\xi^i\}_{i=1,\ldots,\nu}$ of Λ_s, with $\xi^i \in \mathcal{R}$ for any $i = 1, \ldots, \nu$.

To completely characterize $\mathcal{R}$, it is now necessary to have more precise information about η. Pick $\eta \in \mathcal{R}$ such that $2\alpha_n + \eta \in \mathcal{R}$, and decompose it with respect to the $\{\xi^i\}_{i=1,\ldots,\nu}$ basis

$$\eta = \sum_{i=1}^{\nu} q_i \xi^i.$$

The generic form of the superlong roots is thus the following (the form of the other roots being the one given in the previous subsection):

$$\gamma_{sL} = \sum_{i=1}^{\nu} (k_i l_i n_i + q_i)\,\xi^i$$

with $k_i = 1, 2$, $l_i = 1, 2$, n_i running over $\mathbf{Z}$, and q_i being integral numbers, $i = 1, \ldots, \nu$. One is then led to the following situations:

(i) Let $i \in \{1, \ldots, \nu\}$ be such that $k_i l_i = 1$; one can then set, without loss of generality: $q_i = 0$.

(ii) Let $i \in \{1, \ldots, \nu\}$ be such that $k_i l_i = 2$; then

$$2\alpha_n + \eta + 2n\xi^i \in \mathcal{R} \qquad \forall\, n \in \mathbf{Z}$$

One can find $\alpha_L \in \mathcal{R}_l$, such that

$$2\,\frac{(\alpha_L, \alpha_s)}{(\alpha_s, \alpha_s)} = 4\,\frac{(\alpha_L, \alpha_s)}{(\alpha_L, \alpha_L)} = -2$$

Then

$$2\,\frac{(2\alpha_s + \eta + 2n\xi^i, \alpha_L)}{(\alpha_L, \alpha_L)} = -2$$

implies that $2\alpha_s + \alpha_L\eta + 2n\xi^i \in \mathcal{R}\ \forall\, n \in \mathbf{Z}$, and

$$2\,\frac{(2\alpha_s + \alpha_L + \eta + 2n\xi^i, \alpha_L)}{\langle \alpha_s, \alpha_s \rangle} = -2$$

implies that $2\alpha_L + \eta + 2n\xi^i \in \mathcal{R}\ \forall\, n \in \mathbf{Z}$. Then if $k_i = 2$ one can choose $q_i = 0$; if $k_i = 1$ one can choose $q_i = 0$ or 1.

(iii) Let $i \in \{1, \ldots, \nu)$ be such that $k_i l_i = 4$. Then a similar analysis leads to $q_i = 0$ or 2; but these two solutions being equivalent, there is no loss of generality in choosing $q_i = 0$.

It is then time to introduce the following notations:

(i) Let τ be the number of basis vectors ξ^i such that $k_i = 2$ (τ is identical to the twist number previously introduced), $0 \leq \tau \leq \nu$.

(ii) Let μ_1 be the number of basis vectors ξ^i, $i = 1, \ldots, \nu$, such that $k_i = 2$ and $l_i = 2$, $0 \leq \mu_1 \leq \tau$.

(iii) Let μ_2 be the number of basis vectors ξ^i, $i = 1, \ldots, \nu$, such that $k_i = 1$ and $l_i = 2$, $0 < \mu_2 \leq \nu - \tau$.

(It follows from the discussion of the previous subsection that all these twist numbers are well defined).

REMARK. Assume that $q_i = 1$, $i = \tau + 1, \ldots, \sigma$, with $\tau + 1 \leq \sigma \leq \tau + \mu_2$. The generic form of the corresponding part $\Theta^\sigma_{\tau+1}$ of the superlong roots is then

$$\Theta^\sigma_{\tau+1} = \sum_{i=1}^{\sigma} (2n_i + 1)\xi^i$$

It is then possible to exhibit another admissible $\mathbf{Z}$-basis $\{\varepsilon^i\}_{i=1,\ldots,\nu}$ (i.e.,

$\{\varepsilon^i\}_{i=1,\ldots,\nu}$ is a **Z**-basis of Λ_s, $\{k_i\varepsilon^i\}_{i=1,\ldots,\nu}$ is a **Z**-basis of Λ_L, and $\{k_i l_i \varepsilon^i\}_{i=1,\ldots,\nu}$ is a **Z**-basis of Λ_{sL}) such that the corresponding $q_{\tau+1}$ equals 1, and the other corresponding q_i's vanish. One can then assume, without loss of generality, that $q_{\tau+1} = 1$ and $q_i = 0$, $i = 1, \ldots, \nu$, $i \neq \tau + 1$.

The result of the current section can then be summed up as follows.

Theorem 11. Let $\mathcal{R}$ be an irreducible elliptic quasisimple root system of type ν and rank $n + \nu$, and nonreduced gradient root system $\mathcal{R}_{\mathfrak{l}}$. Then $\mathcal{R}$ is completely characterized by $\mathcal{R}_{\mathfrak{l}}$, the type ν, and a set of three non-negative integral "twist numbers" τ, μ_1, μ_2, such that $0 \leq \mu_1 \leq \tau \leq \nu$ and $0 < \mu_2 \leq \nu - \tau$.

The root system $\mathcal{R}$ is then described by

$$\gamma_i = \sum_{i=1}^{\nu} n_i \xi^i$$

$$\gamma_s = \alpha_s + \sum_{i=1}^{\nu} n_i \xi^i$$

$$\gamma_L = \alpha_L + \sum_{i=1}^{\tau} 2n_i \xi^i + \sum_{i=\tau+1}^{\nu} n_i \xi^i$$

$$\gamma_{sL} = 2\alpha_s + \xi^{\tau+1} + \sum_{i=1}^{\mu_1} 4n_i \xi^i + \sum_{i=\mu_1+1}^{\tau+\mu_2} 2n_i \xi^i + \sum_{i=\tau+\mu_2+1}^{\nu} n_i \xi^i$$

where α_s (resp. α_L) stands for a generic short (resp. long) root in $\mathcal{R}_{\mathfrak{l}}$, and n_i, $i = 1, \ldots, \nu$ runs over **Z**. This justifies the notation $\mathcal{R} = (\mathbf{BC}_n; \nu, \tau, \mu_1, \mu_2)$.

REMARK. Putting together Theorems 10 and 11, one obtains a classification of all possible irreducible elliptic quasisimple root systems; however, up to now, such a classification does not extend to a classification of the elliptic quasisimple Lie algebras. For $\nu \leq 1$, this classification is already complete, but for $\nu \leq 2$, one cannot introduce any G.C.M. and apply Chevalley–Harish-Chandra–Serre presentation theorem. The reason is essentially that in these cases, the Weyl group is no longer a Coxeter group [5], but has a more complex structure.

5.4.2 Current Algebra Realizations

As stressed in the previous section, the Chevalley–Harish-Chandra–Serre reconstruction theorem does not apply to elliptic quasisimple root systems of type $\nu \geq 2$. It is then necessary to look for other types of realizations of such root systems, and up to now, the only known realizations are of the "current algebra" type, directly generalizing the loop algebra realizations of affine Kac-Moody algebra.

Throughout this section, one starts from a finite-dimensional simple complex Lie algebra $\mathcal{G}_0$, with Cartan subalgebra $\mathfrak{h}_0$ and root system $\mathcal{R}_0$; denote by $\mathcal{B}_0 = \{\beta_1, \ldots, \beta_n\}$ a basis of simple roots of $\mathcal{G}_0$. Let $\mathcal{L}(\nu) = \mathbf{C}[t_1, t_1^{-1}, \ldots, t_\nu, t_\nu^{-1}]$ be the algebra of Laurent polynomials in the indeterminates $t_1, \ldots, t_\nu$.

The Untwisted Case: $\tau = 0$

The tensor product $\mathcal{G}_1 = \mathcal{L}(\nu) \otimes \mathcal{G}_0$ can be endowed with a Lie structure by pointwise Lie bracket: for any $x_1, x_2 \in \mathcal{G}_0$ and $P_1, P_2 \in \mathcal{L}(\nu)$

$$[P_1 \otimes x_1, P_2 \otimes x_2]_1 = P_1P_2 \otimes [x_1, x_2]_0$$

$\mathcal{G}_1$ is actually not a quasisimple Lie algebra, but can be transformed into an elliptic quasisimple Lie algebra, as follows. Set

$$\mathcal{G}^{\tau=0} = \mathcal{G}_1 \oplus \sum_{i=1}^{\nu}{}^{\oplus} \mathbf{C}c_i \oplus \sum_{i=1}^{\nu}{}^{\oplus} \mathbf{C}d_i$$

$\mathcal{G}^{\tau=0}$ can be endowed with a Lie algebra structure, as follows:

$$\left[(P \otimes x) \oplus \sum_{i=1}^{\nu} \lambda_i c_i \oplus \sum_{i=1}^{\nu} \mu_i d_i, (P' \otimes x') \oplus \sum_{i=1}^{\nu} \lambda_i' c_i \oplus \sum_{i=1}^{\nu} \mu_i' d_i\right]$$

$$= PP' \otimes [x, x']_0 + \sum_{i=1}^{\nu} \mu_i t_i \frac{\partial P'}{\partial t_i} \otimes x'$$

$$- \mu_i' t_i \frac{\partial P}{\partial t_i} \otimes x \oplus \sum_{i=1}^{\nu} \psi_i(P \otimes x, P' \oplus x')c_i$$

with

$$\psi_j(P \otimes x, P' \otimes x') = \frac{1}{(2\pi i)^\nu} (x, x')_0 \oint_0 \frac{\partial P}{\partial t_j} P' \, dt_1 \cdots dt_\nu$$

Such an expression makes sense thanks to the following lemma, whose proof is easy.

Lemma 20. $\psi_i(\cdot, \cdot) \in H^2(\mathcal{G}_1, \mathbf{C})$, $i = 1, \ldots \nu$:

(i) $\psi_i(x, y) = -\psi_i(y, x)\ \forall\, x, y \in \mathcal{G}_1$.

(ii) $\psi_i([x, y], z) + \psi_i([y, z], x) + \psi_i([z, x], y) = 0\ \forall\, x, y, z \in \mathcal{G}_1$.

Define on $\mathcal{G}^{\tau=0}$ the following bilinear form if $x, x' \in \mathcal{G}_0$, $P, P' \in \mathcal{L}(\nu)$, $c, c' \in \Sigma^{\oplus}\, \mathbf{C}c_i$:

$$(P \otimes x \oplus c \oplus d, P' \otimes x' \oplus c' \oplus d')$$

$$= \frac{1}{(2\pi i)^\nu} (x, x')_0 \oint_0 PP' \frac{dt_1}{t_1} \cdots \frac{dt_\nu}{t_\nu} + c \cdot d' + c' \cdot d$$

where the dot stands for the standard component by component scalar product on $\mathbf{C}^\nu$. One then has [8]

Lemma 21. The bilinear form $(\cdot, \cdot)$ is symmetric, invariant and non-degenerate on $\mathcal{G}^{\tau=0}$. It follows that $(\cdot, \cdot)$ is a relevant Killing form for $\mathcal{G}^{\tau=0}$. Note that

$$\mathfrak{h} = \mathfrak{h}_0 \oplus \sum_{i=1}^{\nu}{}^{\oplus} \mathbf{C}c_i \oplus \sum_{i=1}^{\nu}{}^{\oplus} \mathbf{C}d$$

is a Cartan subalgebra of $\mathcal{G}^{\tau=0}$, and has a finite dimension $\dim(\mathfrak{h}) = \mathrm{rk}(\mathcal{G}_0) + 2\nu$.

One can then easily see that $\mathcal{G}^{\tau=0}$ is an elliptic quasisimple Lie algebra. In order to compute its root system, i.e., $\mathrm{Sp}[\mathrm{Ad}(\mathfrak{h})]$, let us introduce $\xi^i \in \mathfrak{h}'$, $i = 1, \ldots, \nu$, by

$$\xi^i(\mathfrak{h}_0) = 0$$
$$\xi^i(c_j) = 0, \qquad j = 1, \ldots, \nu$$
$$\xi^i(d_j) = \delta_{ij}, \qquad j = 1, \ldots, \nu$$

one easily sees that the root system $\mathcal{R}$ of $\mathcal{G}^{\tau=0}$ is of the form

$$\mathcal{R} = \left\{\alpha + \sum_{i=1}^{\nu} n_i\xi^i, \ \alpha \in \mathcal{R}_0, \ n_i \in \mathbf{Z}, \ i = 1, \ldots, \nu\right\}$$

the action of $\mathcal{R}_0 \subset \mathfrak{h}_0$ being trivially extended to $\mathfrak{h}'$:

$$\forall \ \alpha \in \mathcal{R}_0, \ \alpha\left(\sum_{i=1}^{\nu}{}^{\oplus} \mathbf{C}c_i \oplus \sum_{i=1}^{\nu}{}^{\oplus} \mathbf{C}d_i\right) = 0.$$

From this one has the following proposition.

Proposition 13. $\mathcal{G}^{\tau=0}$ is an elliptic quasisimple Lie algebra, whose root system $\mathcal{R}$ has a gradient root system $\mathcal{R}_f = \mathcal{R}_0$, type ν and twist $\tau = 0$: $\mathcal{R} = (\mathcal{R}_0; \nu, 0)$.

REMARK. It is then clear that any untwisted irreducible elliptic quasisimple root system can be realized in this way, as the root system of an untwisted current algebra.

The Twist-One Case: $\tau = 1$

The construction of untwisted current algebras of the previous subsection is a straightforward generalization of the loop algebra construction of affine Lie algebras. This is also the case of the twist-one construction, for which we refer to the subsection of Section 5.2.2 entitled "The Cases $k \neq 1$ (Twisted Cases)." Consider the straight extensions p of the diagram automorphisms ρ_0 of $\mathcal{G}_0$ (of the type $\mathbf{A}_{2l}$, $\mathbf{A}_{2l-1}$, $\mathbf{D}_{l+1}$, $\mathbf{E}_6$, $\mathbf{D}_4$) described there, and form the Lie algebra

$$\mathcal{G}^{\tau=1} = \mathcal{L}(\nu - 1) \otimes \left[\sum_{p \in \mathbf{Z}} \mathbf{c}t_\nu^p \otimes \mathcal{G}_{0;p \bmod k}\right] \oplus \sum_{i=1}^{\nu}{}^{\oplus} \mathbf{C}c_i \oplus \sum_{i=1}^{\nu}{}^{\oplus} \mathbf{C}d_i$$

as a Lie subalgebra of

$$\mathcal{G}^{\tau=0} = \mathcal{L}(\nu) \otimes \mathcal{G}_0 \oplus \sum_{i=1}^{\nu}{}^{\oplus} \mathbf{C}c_i \oplus \sum_{i=1}^{\nu}{}^{\oplus} \mathbf{C}d_i$$

Here, k is the order of ρ and

$$\mathcal{G}_{0;p} = \{g \in \mathcal{G}_0 \text{ s.t. } \rho \cdot g = \varepsilon^p g\}, \qquad \varepsilon = e^{2i\pi/k}$$

Like $\mathcal{G}^{\tau=0}$, $\mathcal{G}^{\tau=1}$ is an elliptic quasisimple Lie algebra; moreover, using Lemma 5, it is easy to derive its root system. The result (a continuation of Proposition 13) is

Proposition 14. $\mathcal{G}^{\tau=1}$ is an elliptic quasisimple Lie algebra, whose root system has a non-simply-laced gradient root system $\mathcal{R}_{\mathfrak{f}}$ (which is nothing but the root system of $\mathcal{G}_{0;0}$ if $\mathcal{G}_0 \neq \mathbf{A}_{2l}$, and is of type $\mathbf{BC}_l$ if $\mathcal{G}_0 = \mathbf{A}_{2l}$), has type ν and twist number $\tau = 1$: $\mathcal{R} = (\mathcal{R}_{\mathfrak{f}}; \nu, 1)$.

The Twist-Two Case, with Reduced $\mathcal{R}_{\mathfrak{f}}$

One is now interested in situations which do not occur in the affine case, i.e., situations in which $\tau = 2$. Let us first describe the constructions leading to twist-two elliptic quasisimple Lie algebras, with reduced gradient root system. The basic observation is that since affine Lie algebras possess a Dynkin diagram, it is easy to build some of their automorphisms. (Of course, a complete study of the automorphisms of Kac Moody algebras is much more complicated; let us quote the work of Levstein [46], who classified the involutive automorphisms of affine Lie algebras.)

The starting point is now an affine Lie algebra $\mathcal{G}_0$, realized as

$$\mathbf{C}[t_\nu, t_\nu^{-1}] \otimes \mathcal{L} \oplus \mathbf{C}c_\nu \oplus \mathbf{C}d_\nu$$

($\mathcal{L}$ being a simple Lie algebra), or a subalgebra of this loop algebra (see Section 5.2.2). Set

$$\overline{\mathcal{G}_0} = \mathcal{G}_0/(\mathbf{C}c_\nu \oplus \mathbf{C}d_\nu)$$

Obviously, the Lie algebra

$$\mathcal{G} = \mathcal{L}(\nu - 1) \otimes \overline{\mathcal{G}_0} \oplus \sum_{i=1}^{\nu}{}^{\oplus} \mathbf{C}c_i \oplus \sum_{i=1}^{\nu}{}^{\oplus} \mathbf{C}d_i$$

(with the usual Lie product and Killing form) is an elliptic quasisimple Lie algebra, and its root system is of the form

$$\mathcal{R} = (\mathcal{R}_{\mathrm{f}}(\mathcal{G}_0), \nu, \tau(\mathcal{G}_0))$$

($\mathcal{R}_{\mathrm{f}}(\mathcal{G}_0)$ being the gradient root system of $\mathcal{G}_0$, and $\tau(\mathcal{G}_0)$ the twist number of $\mathcal{G}_0$) if $\mathcal{R}_{\mathrm{f}}(\mathcal{G}_0)$ is reduced, and

$$\mathcal{R} = (\mathbf{BC}_n; \nu, 0, 0, 1)$$

otherwise.

Let now ρ be the straight extension of an automorphism of the Dynkin diagram of $\mathcal{G}_0$; ρ acts trivially on c_ν and d_ν. Define

$$\mathcal{G}^{\tau=2} = \mathcal{L}(\nu - 2) \otimes \left[\sum_{p\in\mathbf{Z}}{}^{\oplus} \mathbf{C}t_\nu^p \otimes \mathcal{G}_{0;p \bmod k} \right] \oplus \sum_{i=1}^{\nu}{}^{\oplus} \mathbf{C}c_i \oplus \sum_{i=1}^{\nu}{}^{\oplus} \mathbf{C}d_i$$

(k is the order of ρ).

$\mathcal{G}^{\tau=2}$ is a quasisimple Lie algebra, and its root system can be studied in each case. In what follows, we list a family of affine Lie algebras $\mathcal{G}_0$ and their automorphism ρ, and give the root system of the associated $\mathcal{G}^{\tau=2}$. We use the following notations. $\{E_i, F_i, H_i\}_{i=1,\ldots,l}$ are the Chevalley generators of $\mathcal{G}_0$, and $\{\beta_i\}_{i=1,\ldots,l}$ is the corresponding basis of simple roots of $\mathcal{G}_{0;0}$.

(i) $\mathcal{G}_0 = \mathbf{D}_{n+2}^{(1)}$:

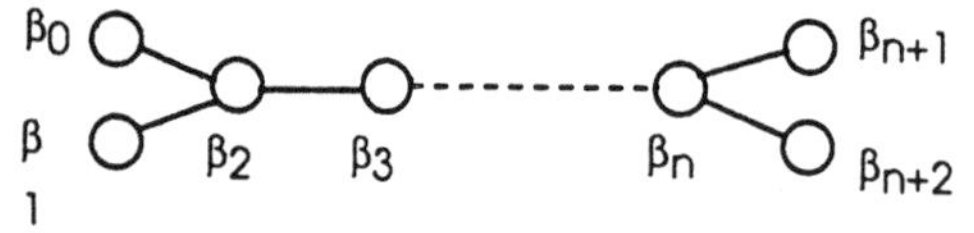

$$\rho(\beta_0) = \beta_1; \quad \rho(\beta_1) = \beta_0; \quad \rho(\beta_{n+1}) = (\beta_{n+2}); \quad \rho(\beta_{n+2}) = (\beta_{n+1})$$

$$\rho(\beta_i) = \beta_i, \qquad i = 2, \ldots, n$$

Then

$$h_0 = H_0 + H_1, \qquad e_0 = E_0 + E_1, \qquad f_0 = F_0 + F_1$$
$$h_i = H_{i+1}, \qquad e_i = E_{i+1}, \qquad f_i = F_{i+1}, \qquad i = 1, \ldots, n$$
$$h_n = H_{n+1} + H_{n+2}, \quad e_n = E_{n+1} + E_{n+2}, \quad f_n = F_{n+1} + F_{n+2}$$

are the Chevalley generators of $\mathcal{G}_{0,0} = \mathbf{D}^{(1)}_{n+2}$, and the root system $\mathcal{R}$ of $\mathcal{G}^{\tau=2}$ is then $\mathcal{R} = (\mathbf{B}_n; \nu, 2)$.

(ii) $\mathcal{G}_0 = \mathbf{D}^{(1)}_{2n}$:

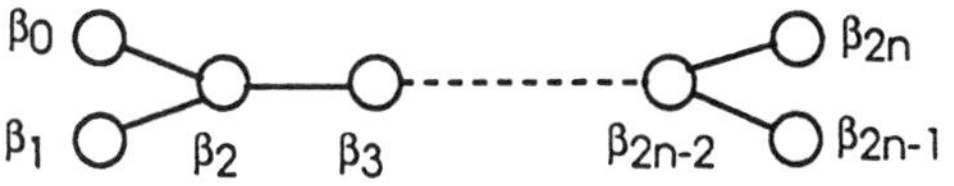

$$\rho(\beta_i) = \beta_{2n-i}, \quad i = 0, \ldots, n-1, n+1, \ldots, 2n; \qquad \rho(\beta_n) = \beta_n$$

Then the following elements

$$h_i = H_i + H_{2n-i}, \quad e_i = E_i + E_{2n-i}, \quad f_i = F_i + F_{2n-i}, \quad i = 0, \ldots n-1$$
$$h_n = H_n, \qquad e_n = E_n, \qquad f_n = F_n$$

form a set of Chevalley generators of $\mathcal{G}_{0,0} = \mathbf{A}^{(2)}_{2n-1}$, and the root system $\mathcal{R}$ of $\mathcal{G}^{\tau=2}$ is then $\mathcal{R} = (\mathbf{C}_n; \nu, 2)$.

(iii) $\mathcal{G}_{0,0} = \mathbf{E}^{(1)}_6$:

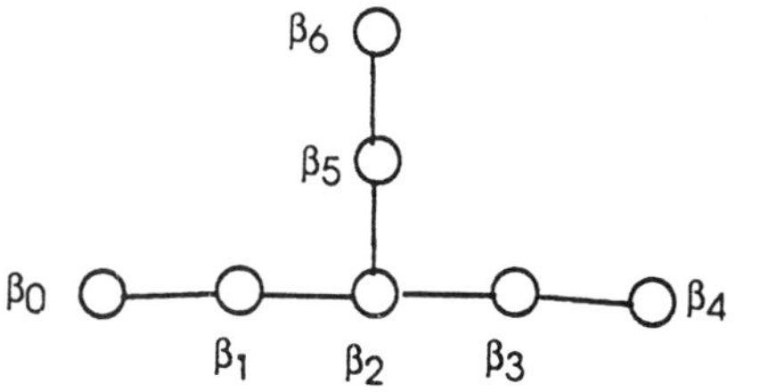

$$\rho(\beta_0) = \beta_6; \quad \rho(\beta_6) = \beta_4; \quad \rho(\beta_4) = \beta_0$$
$$\rho(\beta_1) = \beta_5; \quad \rho(\beta_5) = \beta_3; \quad \rho(\beta_3) = \beta_1$$
$$\rho(\beta_2) = \beta_2$$

Then the elements

$$\begin{aligned} h_0 &= H_0 + H_6 + H_4, & e_0 &= E_0 + E_6 + E_4, & f_0 &= F_0 + F_6 + F_4 \\ h_1 &= H_1 + H_5 + H_3, & e_1 &= E_1 + E_5 + E_3, & f_1 &= F_1 + F_5 + F_3 \\ h_2 &= H_2 & e_2 &= E_2, & f_2 &= F_2 \end{aligned}$$

are the Chevalley generators of $\mathcal{G}_{0,0} = \mathbf{D}_4^{(3)}$, and the root system $\mathcal{R}$ of $\mathcal{G}^{\tau=2}$ is $\mathcal{R} = (\mathbf{G}_2; \nu, 2)$.

(iv) $\mathcal{G}_0 = \mathbf{E}_7^{(1)}$:

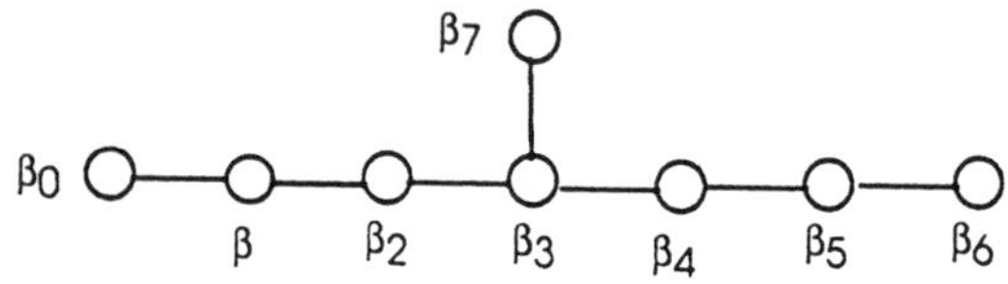

$$\rho(\beta_i) = \beta_{6-i}, \quad i = 0, \ldots, 6: \qquad \rho(\beta_7) = \beta_7$$

Then

$$\begin{aligned} h_i &= H_i + H_{6-i}, & e_i &= E_i + E_{6-i}, & f_i &= F_i + F_{6-i}, \qquad i = 0, 1, 2 \\ h_3 &= H_3, & e_3 &= E_3, & f_3 &= F_3 \\ h_4 &= H_7, & e_4 &= E_7, & f_4 &= F_7 \end{aligned}$$

are the Chevalley generators of $\mathcal{G}_{0,0} = \mathbf{E}_6^{(2)}$, and the root system $\mathcal{R}$ of $\mathcal{G}^{\tau=2}$ is $\mathcal{R} = (\mathbf{F}_4; \nu, 2)$. Summarizing, we have

Proposition 15. $\mathcal{G}^{\tau=2}$ is an elliptic quasisimple Lie algebra, whose root system has a non-simply-laced reduced gradient root system $\mathcal{R}_{\text{f}}$, has type ν and twist number $\tau = 2$: $\mathcal{R} = (\mathcal{R}_{\text{f}}; \nu, 2)$.

The Twist-Two Case, with Nonreduced $\mathcal{R}_{\text{f}}$

This subsection is concerned with the same construction as in the previous one, leading now to quasisimple root systems with nonreduced gradient root systems. With the same notations as in the previous subsection set

$$\mathcal{G}_{\text{NR}}^{\tau=2} = \mathcal{L}(\nu - 2) \otimes \left[\sum_{p \in \mathbf{Z}}^{\oplus} \mathbf{C}t_{\nu-1}^{p} \otimes \mathcal{G}_{0;p \bmod k} \right] \oplus \sum_{i=1}^{\nu}{}^{\oplus}\, \mathbf{C}c_i \oplus \sum_{i=1}^{\nu}{}^{\oplus}\, \mathbf{C}d_i$$

in the following cases.

(i) $\mathcal{G}_0 = \mathbf{A}^{(1)}_{2n+1}$:

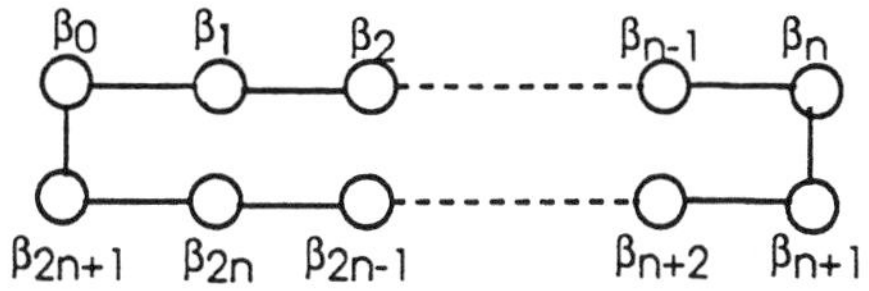

$$\rho(\beta_i) = \beta_{2n+1-i}, \qquad i = 0, \ldots, 2n+1$$

Then the elements

$$h_0 = 2(H_0 + H_{2n+1}), \quad e_0 = 2(E_0 + E_{2n+1}), \quad f_0 = F_0 + F_{2n+1}$$
$$h_n = 2(H_n + H_{n+1}), \quad e_n = 2(E_n + E_{n+1}), \quad f_n = F_n + F_{n+1}$$
$$h_i = (H_i + H_{2n+1-i}), \quad e_i = (E_i + E_{2n+1-i}), \quad f_i = F_i + F_{2n+1-i},$$
$$i = 1, \ldots, n-1$$

are the Chevalley generators of $\mathcal{G}_{0,0} = \mathbf{D}^{(2)}_{n+1}$, and an explicit computation of the root system $\mathcal{R}$ of $\mathcal{G}^{\tau=2}_{\mathrm{NR}}$ shows that $\mathcal{R} = (\mathbf{BC}_n;\ \nu,\ 1,\ 0,\ 1)$.

(ii) $\mathcal{G}_0 = \mathbf{D}^{(2)}_{2n+1}$:

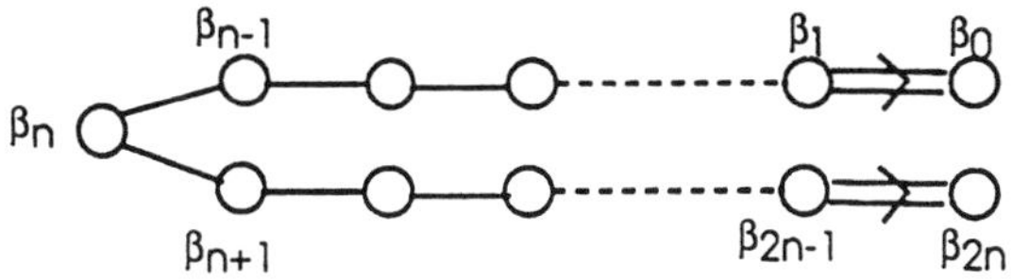

$$\rho(\beta_i) = \beta_{2n-i}, \qquad i = 0, \ldots, 2n$$

Then the elements

$$h_i = 2(H_i + H_{2n-i}), \quad e_i = 2(E_i + E_{2n-i}), \quad f_i = F_i + F_{2n-i},$$
$$i = 0, \ldots, 2n$$
$$h_n = H_n, \qquad e_n = E_n, \qquad f_n = F_n$$

are the Chevalley generators of $\mathcal{G}_{0,0} = \mathbf{A}^{(2)}_{2n}$, and an explicit computation of the root system $\mathcal{R}$ of $\mathcal{G}^{\tau=2}_{\mathrm{NR}}$ shows that $\mathcal{R} = (\mathbf{BC}_n;\ \nu,\ 0,\ 0,\ 2)$.

(iii) $\mathcal{G}_0 = \mathbf{D}^{(1)}_{2n+2}$:

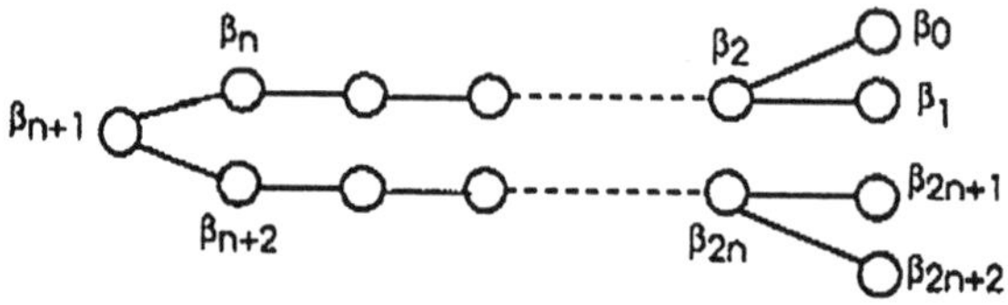

$$\rho(\beta_i) = \beta_{2n+2-i}, \qquad i = 2, \ldots, 2n$$

$$\rho(\beta_0) = \beta_{2n+1}, \quad \rho(\beta_{2n+1}) = \beta_1, \quad \rho(\beta_1) = \beta_{2n+2}, \quad \rho(\beta_{2n+2}) = \beta_0$$

Then the elements

$$\begin{aligned}
h_0 &= H_0 + H_1 + H_{2n+1} + H_{2n+2},\\
e_0 &= E_0 + E_1 + E_{2n+1} + E_{2n+2},\\
f_0 &= F_0 + F_1 + F_{2n+1} + F_{2n+2}\\
h_i &= H_{i+1} + H_{2n+1-i},\\
e_i &= E_{i+1} + E_{2n+1-i},\\
f_i &= F_{i+1} + F_{2n+1-i}, \qquad i = 1, \ldots, n-1\\
h_n &= H_{n+1}\\
e_n &= E_{n+1}\\
f_n &= F_{n+1}
\end{aligned}$$

are the Chevalley generators of $\mathcal{G}_{0,0} = \mathbf{A}^{(2)}_{2n}$, and an explicit computation of the root system $\mathcal{R}$ of $\mathcal{G}^{\tau=2}_{\mathrm{NR}}$ shows that $\mathcal{R} = (\mathbf{BC}_n; \nu, 1, 1, 1)$.

Summarizing, one then has the following result.

Proposition 16. $\mathcal{G}^{\tau=2}_{\mathrm{NR}}$ is an elliptic quasisimple Lie algebra, with non-reduced gradient root system, and its root system takes the form $\mathcal{R} = (\mathbf{BC}_n; \nu, \tau, \mu_1, \mu_2)$, where τ, μ_1, μ_2 exhaust all their possible values when $\nu = 2$.

Conclusions

The constructions presented in Section 5.4.2 and more specially Propositions 13 to 16 have the following corollary.

Corollary. All the elliptic quasisimple root systems with type $\nu \le 2$ can be realized as the root systems of current algebras, as constructed in the current section.

REMARKS. It is theoretically possible to look for twist-three current algebras, starting from automorphisms of twist-two algebras, however, this would require a more precise knowledge of such automorphisms.

Although no explicit realization has been found, we believe that it is possible to associate a quasisimple Lie algebra to any elliptic quasisimple root system classified in Theorems 10 and 11; however, it is also possible that for $\nu > 2$, a given elliptic quasisimple root system is not the root system of a unique quasisimple Lie algebra, but of many of them. This remains an open problem.

In the $\nu = 1$ case, $\mathcal{G}$ inherits a canonical projective action of the Virasoro algebra, as discussed previously. In the $\nu > 1$ cases, the Lie algebra which appears is the Lie algebra of polynomial vector fields on the ν-dimensional torus $\mathbf{T}^\nu$, with generators $\{\mathcal{L}^{(i)}_{n_1,\ldots,n_\nu}, i = 1, \ldots, \nu, n_i \in \mathbf{Z}\}$ and relations

$$[\mathcal{L}^{(i)}_{n_1,\ldots,n_\nu}, \mathcal{L}^{(j)}_{m_1,\ldots,m_\nu}] = m_i \mathcal{L}^{(j)}_{m_1+n_1,\ldots,m_\nu+n_1} - n_j \mathcal{L}^{(i)}_{m_1+n_1,\ldots,m_\nu+n_1}$$

In the $\nu = 2$ case, this Lie algebra recently appeared in the study of membrane theories.

5.4.3 Real Quasisimple Lie Algebras

As usual, the real forms of any quasisimple Lie algebra $\mathcal{G}$ are defined by a consistent conjugation ω on $\mathcal{G}$ (see Sections 5.2.1 and 5.2.3 for precise definitions). Recall that the real form associated with the conjugation ω is then defined by

$$\mathcal{G}_\omega = \{g \in \mathcal{G} \text{ s.t. } \omega \cdot g = -g\}.$$

Here, one will be essentially interested in the compact real forms of (untwisted or twisted) current algebras. It is then necessary to introduce the corresponding Cartan involution with the notations of the previous section, let

$$\mathcal{G} = \mathcal{L}(\nu) \otimes \mathcal{G}_0 \quad \oplus \sum_{i=1}^{\nu}{}^{\oplus} \mathbf{C}c_i \oplus \sum_{i=1}^{\nu}{}^{\oplus} \mathbf{C}d_i$$

be any untwisted current algebra of type ν. As a direct generalization of the affine case, the Cartan conjugation is defined by its action on elements of the form $P(t_1, \ldots, t_\nu) \otimes x \in \mathcal{L}(\nu) \otimes \mathcal{G}_0$, by

$$\omega_c \cdot P(t_1, \ldots, t_\nu) \otimes x = \overline{P(t_1^{-1}, \ldots, t_\nu^{-1})} \otimes \omega_0(x)$$

ω_0 being the Cartan conjugation on $\mathcal{G}_0$, and the bar denoting complex conjugation.

The definition of ω_c is then completed by setting

$$\omega_c(c_i) = -c_i, \qquad i = 1, \ldots, \nu$$

$$\omega_c(d_i) = -d_i, \qquad i = 1, \ldots, \nu$$

Under these conditions, the compact real form $\mathcal{G}_c$ of $\mathcal{G}$ is naturally identified with the Lie algebra

$$\mathcal{G}_c \cong P(\mathbf{T}^\nu, \mathcal{G}_{0c}) \oplus \sum_{i=1}^{\nu}{}^{\oplus} \mathbf{R}c_i \oplus \sum_{i=1}^{\nu}{}^{\oplus} \mathbf{R}d_i$$

$P(\mathbf{T}^\nu, \mathcal{G}_{0c})$ being the Lie algebra of polynomial mappings from the ν-dimensional torus $\mathbf{T}^\nu$ (set of ν-tuples of complex numbers of modulus equal to 1) into, $\mathcal{G}_{0c}$, with pointwise Lie bracket. This explains the terminology "current algebra" or "generalized loop algebra" [67] or "gauge algebra" [27].

5.4.4 Highest Weight Representations of Compact Current Algebras

We restrict ourselves here to the case of unitary highest weight representations of the compact real form of current algebras, which we will assume untwisted, for simplicity (the result is similar in the twisted case). Let then

$$\mathcal{G} = \mathcal{L}(\nu) \otimes \mathcal{G}_0 \quad \oplus \sum_{i=1}^{\nu}{}^{\oplus} \mathbf{C}c_i \oplus \sum_{i=1}^{\nu}{}^{\oplus} \mathbf{C}d_i$$

be any untwisted current algebra, and let $\mathcal{G}_c$ be its compact real form. We will closely follow the description of Section 5.2.3 in the integrable case. To build a Verma module, the first thing one needs is a parabolic subalgebra $\mathfrak{p}$ of $\mathcal{G}$, consistent with the Cartan conjugation ω_c. In affine case, such a subalgebra can easily be constructed from the possible triangular decompositions of $\mathcal{G}$. Conversely, for $\nu \geq 2$, the impossibility of building a canonical total ordering on the root system $\mathcal{R}$ implies that there is no canonical way of constructing such a parabolic subalgebra. It is then necessary to introduce the following notion of projective ordering, or ε-*ordering* [67]:

Definition 9. Let $\{\varepsilon_1, \ldots, \varepsilon_\nu\} = \varepsilon$ be any family of ν real numbers; a root $\gamma = \alpha + \Sigma_{i=1}^{\nu} m_i\xi^i$ of $\mathcal{G}$ is said to be ε-*positive* if either $\Sigma_{i=1}^{\nu} m_i\varepsilon^i > 0$, or $\Sigma_{i=1}^{\nu} m_i\varepsilon^i = 0$ and $\alpha > 0$ with respect to the standard ordering of $\mathcal{G}_0$.

REMARK. If $\nu = 0$, or $\nu = 1$ and $\varepsilon_1 \neq 0$, this ordering is nothing but the standard ordering defined by the G.C.M. of $\mathcal{G}$ (up to a sign).

If $\nu = 1$ and $\varepsilon_1 = 0$, the ε-ordering coincides with the natural ordering described in Section 5.2.3 (see also [29]).

The ε-ordering is not able to reproduce the mixed orderings introduced in Section 5.2.3; the construction presented below does not then apply to the exceptional representations.

In the general case, with respect to the ε-ordering, the root system $\mathcal{R}$ decomposes as $\mathcal{G} = \mathfrak{n}_- \oplus \mathfrak{h} \oplus \mathfrak{n}_0 \oplus \mathfrak{n}_+$ (with standard notations), including the following decomposition of the universal envelope of $\mathcal{G}$ [9]:

$$\mathcal{U}(\mathcal{G}) = \mathcal{U}(\mathfrak{n}_-) \otimes \mathcal{U}(\mathfrak{h}) \otimes \mathcal{U}(\mathfrak{n}_0) \otimes \mathcal{U}(\mathfrak{n}_+)$$

Moreover, setting $\mathfrak{p} = \mathfrak{h} \oplus \mathfrak{n}_0 \oplus \mathfrak{n}_+$, one has

$$\mathcal{U}(\mathcal{G}) = \mathfrak{n}_-\mathcal{U}(\mathcal{G}) \otimes \mathcal{U}(\mathfrak{p})$$

Let P be the orthogonal projector on $\mathcal{U}(\mathfrak{p})$ associated with this decomposition, and let Λ be a complex character of $\mathfrak{p}$, such that

$$\Lambda(P(\omega \cdot v)) = \Lambda(P \cdot v)^*, \qquad \forall\, v \in \mathcal{U}(\mathcal{G})$$

one can then perform Verma's construction [71] following Section 5.2.3; introduce [67]

$$\begin{aligned}
\mathcal{I}(\Lambda) &= \mathcal{U}(\mathcal{G})\{p - \Lambda(p),\, p \in \mathfrak{p}\}\\
\mathrm{M}(\Lambda) &= \mathcal{U}(\mathcal{G})/\mathcal{I}(\Lambda)\\
\mathcal{H}_1(u, v) &= \Lambda(P((\omega \cdot v)u)) \ \forall\, u, v \in M(\Lambda)\\
R(\Lambda) &= \{u \in M(\Lambda),\, \mathcal{H}_1(u, v) = 0,\ \forall\, v \in M(\Lambda)\}\\
L(\Lambda) &= M(\Lambda)/R(\Lambda)\\
\mathcal{H} &= \mathcal{H}_1|_{L(\Lambda)}
\end{aligned}$$

As usual, the highest weight vector v_Λ is the canonical image of $1 \in \mathcal{U}(\mathcal{G})$ in this quotient.

For simplicity, it is convenient to introduce the following notation. Let

$$e_i = t \otimes E_i, \quad f_i = t^{-1} \otimes F_i, \quad h_i = 1 \otimes H_i$$

where $\{E_i, F_i, H_i, i = 1, \ldots, n\}$ is a set of Chevalley generators of $\mathcal{G}_0$, and let $e^i \in \mathcal{G}_{\theta+\xi^i}$, $f^i \in \mathcal{G}_{\theta+\xi^i}$ and $h^i \in \mathfrak{h}$, normalized so that $[e^i, f^i] = h^i[h^i, e^i] = 2e^i[h^i, f^i] = -2f^i$ (note that $[e^i, f^i] \neq 0$) (θ is here the highest root). As in Section 5.2.3, the module $L(\Lambda)$ is said to be of the nilpotent type if for any $x \in \{e_1, \ldots, e_n, e^1, \ldots e^\nu, f_1, \ldots, f_n, f^1, \ldots, f^\nu\}$, x acts on $L(\Lambda)$ as a locally nilpotent operator. Moreover, it is easy to see that a necessary and sufficient condition for such an x to be nilpotent is that there exists $m \in \mathbb{N}$ such that $x^m \cdot v_\Lambda = 0$ (see [66] for instance). From

$$e_i f_i^{n+1} \cdot v_\Lambda = (n + 1)(\Lambda(h_i) - n)e_i f_i^n v_\Lambda$$

and

$$e_j(f^j)^{n+1} \cdot v_\Lambda = (n + 1)(\Lambda(h^j) - n)e^j(f^j)^n \cdot v_\Lambda,$$

it follows easily that the following lemma holds.

Lemma 22. $L(\Lambda)$ is of the nilpotent type if and only if $\Lambda(h_i) \in \mathbb{N}$, $i = 1, \ldots, n$, and $\Lambda(h_j) \in \mathbb{N}$, $j = 1, \ldots, \nu$, assuming that all ξ^i are positive isotropic roots. When studying the action on $L(\Lambda)$ of the center $\Sigma_{i=1}^{\oplus\nu} \mathbf{C}c_i$ of $\mathcal{G}$, it turns out that the notion of positivity of an arbitrary isotropic root $\delta = \Sigma_{i=1}^{\oplus\nu} m_i\xi^i$ is actually given by the positivity of the product $(\delta, \Lambda) = \Sigma_{i=1}^{\nu} m_i(\xi^i, \Lambda)$. This forces the ε_i-numbers appearing in the projective ordering to be proportional to the numbers $\mu_i = (\Lambda, \xi^i)$ $i = 1, \ldots, \nu$. For convenience, let us choose

$$\varepsilon_i = 2\frac{(\Lambda, \xi^i)}{(\theta, \theta)}, \qquad i = 1, \ldots, \nu$$

Lemma 22 has then the immediate following corollary.

Corollary. $L(\Lambda)$ is of the nilpotent type if and only if $\Lambda(h_i) \in \mathbb{N}$, $i = 1, \ldots, n$, and $\varepsilon_j \in \mathbb{N}$, with $\varepsilon_j = 0$ or $\varepsilon_j \geq 2(\Lambda, \theta)/(\theta, \theta)$.

PROOF. It suffices to compute

$$\Lambda(h^j) = -2\frac{(\Lambda, \theta)}{(\theta, \theta)} + \varepsilon_j, \qquad j = 1, \ldots, \nu,$$

if $\varepsilon_j = 0$, $-\theta + \xi^j$ is not a positive root, and if $\varepsilon_j \neq 0$, apply directly Lemma 22. □

It is now time to apply the results of Section 5.2.3 to get a classification of the unitary highest weight $\mathcal{G}_c$-modules. Consider first the "natural" case, i.e., the case where $\varepsilon_i = 0$, $i = 1, \ldots, \nu$. Then

$$\mathfrak{p} = \mathfrak{h} \oplus \mathfrak{n}_0 \oplus \mathfrak{n}_+$$

$$\mathfrak{n}_0 = \sum_{\{m\}\in\mathbf{Z}^\nu}^{\oplus} \mathcal{G}_{\Sigma_{i=1}^{\oplus\nu} m_i\xi^i}$$

$$\mathfrak{n}_+ = \sum_{\substack{\{m\}\in\mathbf{Z}^\nu \\ \alpha\in(\mathcal{R}_0)_+}}^{\oplus} \mathcal{G}_{\alpha + \Sigma_{i=1}^{\oplus\nu} m_i\xi^i}$$

One defines a $\mathcal{G}$-module as follows: let $\{c_i^j; j = 1, \ldots, \nu, i = 1, \ldots, N, N \in \mathbb{N}\}$ be a family of real numbers, and define the projections: ϕ_i

$\mathcal{G} \to \mathcal{G}_0$, $i = 0, \ldots, N$:

$$t^m \otimes g \to \phi_i(t_1^{m_1} \ldots, t_\nu^{m_\nu} \otimes g) = \prod_{j=1}^{\nu} (c_i^j)^{m_j} \cdot g, \qquad g \in \mathcal{G}_0, \quad m \in \mathbf{Z}_\nu$$

$$c_j \to \phi_i(c_j) = 0, \qquad j = 1, \ldots, \nu$$

$$d_j \to \phi_i(d_j) = 0, \qquad j = 1, \ldots, \nu$$

Let now $\pi_0^1, \ldots, \pi_0^N$ be irreducible unitary highest weight representations of $\mathcal{G}_0$, with respective highest weights $\lambda_0^1, \ldots, \lambda_0^N$. This allows us to introduce the following character of $\mathfrak{p}$:

$$\Lambda : \mathfrak{p} \to \mathbf{C}$$

$$t_1^{m_1} \ldots t_\nu^{m_\nu} \otimes p_0 \to \Lambda(t_1^{m_1} \ldots t_\nu^{m_\nu} \otimes p_0) = \left[\sum_{i=1}^{N} \prod_{j=1}^{\nu} (c_i^j)^{m_j} \right] \lambda_0^i(p_0)$$

$$c_i \to \Lambda(c_i) = 0, \qquad i = 1, \ldots, \nu$$

$$d_i \to \Lambda(d_i) = 0, \qquad i = 1, \ldots, \nu$$

According to the previous construction, one then gets the following representation π of $\mathcal{G}$ on $L(\Lambda)$:

$$L(\Lambda) = L(\lambda_0^1) \otimes L(\lambda_0^2) \otimes \cdots \otimes L(\lambda_0^N)$$

$$\pi = (\pi_0^1 \circ \phi_1) + (\pi_0^2 \circ \phi_2) + \cdots + (\pi_0^N \circ \phi_N)$$

which is obviously unitarisable, but far from being irreducible in general. Such representations of affine Lie algebras, are called *elementary* too (or *located*).

Consider now the case where some of the ε_i's are nonzero. Then, the projector involved in the ε-ordering has $(\nu - 1)$-dimensional kernel, and there is no loss of generality in assuming $\varepsilon \equiv \varepsilon_1 \neq 0$, and $\varepsilon_2 = \cdots = \varepsilon_\nu = 0$. One then build a $\mathcal{G}$-module $(\pi, L(\Lambda))$ of highest weight Λ as follows: let $\{d_i^j, i = 1, \ldots, N, j = 2, \ldots, \nu\}$ be any family of real numbers, and introduce the following evaluation functionals:

$$\psi_i; \mathcal{G} \to \mathcal{G}_1 = \mathbf{C}(t_1, t_1^{-1}) \otimes \mathcal{G}_0 \oplus \mathbf{C}c_1 \oplus \mathbf{C}d_1$$

$$t_1^{m_1} \ldots t_\nu^{m_\nu} \otimes p_0 \to \psi_i(t_1^{m_1} \ldots t_\nu^{m_\nu} \otimes g) = \prod_{j=2}^{\nu} (d_i^j)^{m_j} t_1^{m_1} \otimes \mathcal{G}$$

$$c_1 \to \psi_i(c_1) = c_1$$

$$c_j \to \psi_i(c_j) = 0, \qquad j = 2, \ldots, \nu$$

$$d_1 \to \psi_i(d_1) = d_1$$

$$d_j \to \psi_i(d_j) = 0, \qquad j = 2, \ldots, \nu$$

The parabolic $\mathcal{G}$-subalgebra considered here is of the form

$$\mathfrak{p} = \mathfrak{h} \oplus \mathfrak{n}_0 \oplus \mathfrak{n}_+$$

with

$$\mathfrak{n}_0 = \sum_{\{m\}\in \mathbf{Z}^{\nu-1}}^{\oplus} \mathcal{G}_{\sum_{i=2}^{\nu}\oplus m_i\xi^i} \qquad n_+ = \sum_{\substack{\{m\}\in \mathbf{Z}^{\nu} \\ m_1>0 \\ \alpha\in\mathcal{R}_0}}^{\oplus} \mathcal{G}_{\alpha+\sum_{i=1}^{\nu}\oplus m_i\xi^i}$$

Let $\{\pi_1^1, \ldots, \pi_1^N\}$ be a set of integrable representations of $\mathcal{G}_1$ on $L(\lambda_1^1), \ldots, L(\lambda_1^N)$ respectively. Introduce the following character $\Lambda \in \mathfrak{p}$, defined by

$$\Lambda(t_1^{m_1} \ldots t_\nu^{m_\nu} \otimes p_0) = \sum_{j=1}^{\nu} \left[\prod_{j=1}^{\nu} (d_i^j)^{m_j} \right] \lambda_1^i(t_1^{m_1} \otimes p_0)$$

$$\Lambda(c_i) = \varepsilon$$

$$\Lambda(c_i) = 0, \qquad j = 2, \ldots, \nu$$

Let us then build the corresponding Verma module $L(\Lambda)$, which is of the form

$$L(\Lambda) = L(\lambda_1^1) \otimes L(\lambda_1^2) \otimes \cdots \otimes L(\lambda_1^N)$$

$$\pi = (\pi_1^1 \circ \psi_1) + (\pi_1^2 \circ \psi_2) + \cdots + (\pi_1^N \circ \psi_n)$$

From the results of Section 5.2.3, $L(\Lambda)$ is obviously unitarizable, but in general far from being irreducible. The $\mathcal{G}$-modules of this type are called *semielementary* (or *semilocated*). Taking into account the description given in Section 5.2.3, one can then state the following.

Theorem 12. Let $\mathcal{G}$ be any untwisted current algebra of type ν. Then the only possible unitary highest weight representations of the compact real form $\mathcal{G}_0$ of $\mathcal{G}$ are either elementary or semielementary.

Corollary. The gauge algebra $\mathcal{G}_c = P(\mathbf{T}^\nu, \mathcal{G}_{0c})$ does not admit linear unitary highest weight representations apart from the located representations. The semilocated representations are projective representations, and are not faithful for $\nu > 1$.

REMARKS. 1. Theorem 12 holds as well for twisted current algebras constructed in Section 5.4.2, considered as subalgebras of the untwisted ones.

It turns out that the strategic argument in the above discussion is the following take any $\mathbf{T}^1$-subgroup of $\mathbf{T}^\nu$, such that the corresponding "central charge" (i.e., the corresponding linear combination of the $\Lambda(c_j)$'s vanishes). Then, from the results of Section 5.2.3, the corresponding affine $\mathcal{G}$-subalgebra $P(\mathbf{T}^1, \mathcal{G}_0) \oplus \mathbf{C}c \oplus \mathbf{C}d$ is elementarily represented.

An alternative to this construction would be to consider unitary representations of $\mathcal{G}_\omega$, ω being the noncompact conjugation on $\mathcal{G}$ corresponding to $\mathcal{G}_\omega = P(\mathbf{T}^\nu, su(n, 1))$. From the results of Section 5.2.3, the central charge is then allowed to vanish, and one should then obtain the generalizations of exceptional representations of $P(\mathbf{T}^1, su(n, 1))$ [29] such representations also seem to be strongly connected to the "continuous tens or product" representations of noncompact current groups [8] $\mathfrak{D}(M, SU(n, 1))$ (see Chapter 2).

REFERENCES

1. V. Alessandrini, P. Binetruy, C. Kounnas, and A. Schwimmer, in *Théories de Cordes et Supercordes*, Proceedings of the GIF Summer School, Marseille (1987).
2. R. C. Arcuri, Ph.D. thesis, Imperial College (1989).
3. F. A. Bais, P. Bouwknegt, M. Surridge, and K. Schoutens, *Nucl. Phys. B,304*: 348 (1988).
4. F. A. Bais, P. Bouwknegt, M. Surridge, and K. Schoutens, *Nucl. Phys. B,304*: 371 (1988).
5. N. Bourbaki, *Groupes et Algèbres de Lie*, Masson, Paris, Chaps. 4, 5, 6(1981).
6. E. Cartan, Thesis (1894).
7. C. Chevalley, *C.R.A.S.*, *227*: 1136 (1948).
8. P. Delorme, Private communication.
9. J. Dixmier, *Algèbres Enveloppantes*, Gauthier-Villars, Paris (1974).
10. V. Drinfeld, in Proceedings of the Conference "Infinite-Dimensional Lie Algebras," Ed. V. Kac, World Scient., Singapore (1989) Marseille (1988).
11. V. A. Fateev, A. B. Zamolodchikov, *Nucl. Phys. B,280*: 644 (1987).
12. B. L. Feigin and D. B. Fuchs, *Funct. Anal. Appl.*, *17*: 91 (1983).
13. I. B. Frenkel, *Beyond Affine Lie Algebras*, in Proceedings of the International Conference of Mathematicians, Berkeley, 1986, vol. 1, American Mathematical Society, pp. 821–839 (1987).
14. I. B. Frenkel and V. G. Kac, *Invent. Math.*, *62*: 23 (1980).
15. I. B. Frenkel, J. Lepowsky, and A. Meurman, in *Mathematical Aspects of String Theory* (S. T. Yau, ed.), San Diego World Scient., Singapore (1987) (1988).
16. I. Frenkel, J. Lepowsky, and A. Meurman, *Vertex Operator Algebras and the Monster*, Academic Press, New York (1988).
17. H. Freudenthal and H. De Vries, *Linear Lie Groups*, Academic Press, New York (1969).
18. D. Friedan, Z. Qiu, and S. Shenker, Proceedings of the conference *Vertex Operators in Mathematics and Physics*, (J. Lepowsky, ed.), MSRI, vol. 3, Springer, New York (1985).

19. H. Garland, *Publ. Math. IHES 52*: 5 (1980).
20. I. M. Gelfand, *Actes, Congrès Int. Math.*, *1*: 95 (1970).
21. P. Goddard, A. Kent, and D. Olive, *Phys. Lett. 152B*: 88 (1985).
22. P. Goddard, A. Kent, and D. Olive, *Comm. Math. Phys.*, *103*: 105 (1986).
23. P. Goddard and D. Olive, *Int. J. Mod. Phys. A,1*: 303 (1986).
24. L. V. Goncharova, *Funct. Anal. Prilozhen*, *7*, no. 2: 6; no. 3: 33 (1973).
25. M. B. Green, J. H. Schwarz, and E. Witten, *Superstring Theory*, Cambridge University Press, Cambridge (1978).
26. Harish-Chandra, *Trans. Amer. Math. Soc. 70*: 28 (1951).
27. R. Høegh-Krohn and B. Torresani, *J. Funct. Anal.*, *89*: 106 (1990).
28. H. P. Jakobsen, *J. Funct. Anal.*, *52*: 385 (1983).
29. H. P. Jakobsen and V. G. Kac, *Non-linear Equation in Quantum Field Theory*, Lecture Notes in Physics, vol. 226, Springer-Verlag, Berlin and New York, pp. 1–20 (1985).
30. J. C. Jantzen, *Einhüllende Algebren halbeinfacher Lie-Algebren*, Springer-Verlag, Berlin (1983).
31. V. G. Kac, *Funct. Anal. Appl.*, *1*: 328 (1967) (transl.).
32. V. G. Kac, *Math. USSR Izv.*, *2*: 1271 (1968) (transl.).
33. V. G. Kac, *Funct. Anal. Appl.*, *8*: 68 (1974) (transl.).
34. V. G. Kac, *Highest Weight Representations of Infinite-Dimensional Lie Algebras*, Proceedings of the International Congress of Mathematicians, Helsinki, 1978, Acad. Scientiarum Fennica, Helsinki, pp. 299–304 (1980).
35. V. G. Kac, *Infinite-Dimensional Lie algebras*, 2nd ed., Cambridge University Press (1985).
36. V. G. Kac, D. A. Kazhdan, J. Lepowsky, and R. L. Wilson, *Adv. Math.*, *42*: 83 (1981).
37. V. G. Kac and D. H. Peterson, in *Arithmetics and Geometry* (M. Artin and J. Tate, eds.), Birkhäuser, Boston, pp. 141–166 (1983).
38. V. G. Kac and D. H. Peterson, *Adv. Math.*, *53*: 125 (1984).
39. V. G. Kac and D. H. Peterson, *112 Constructions of the Basic Representation of the Loop Group of* E_8, Proceedings of the Conference on Anomalies, Geometry, Topology, Argonne, Eds. W. A. Bardeen, A. R. White, World Scientific, Singapore, pp. 176–298 (1985).
40. V. G. Kac and A. K. Raina, *Bombay Lectures on Highest Weight Representations of Infinite-Dimensional Lie Algebras*, World Scientific, Singapore (1987).
41. V. G. Kac and I. T. Todorov, *Comm. Math. Phys.*, *102*: 337 (1985).
42. V. G. Kac and M. Wakimoto, *Unitarisable Highest Weight Representations of the Virasoro, Neveu-Schwarz and Ramond Algebras*, in Lecture Notes in Physics, vol. 261, Springer-Verlag, Berlin and New York (1988).
43. V. G. Kac and M. Wakimoto, *Adv. Math.*, *70*: 156 (1988).
44. J. Lepowsky, *Proc. Nat. Acad. Sci. USA*, *82*: 8295 (1985).
45. J. Lepowsky and R. L. Wilson, *Comm. Math. Phys.*, *62*: 43 (1978).
46. F. Levstein, M.I.T. dissertation (1984).
47. R. Marcuson, *J. Algebra*, *34*: 84 (1975).

48. I. G. McDonald, *Invent. Math.*, *15*: 91 (1972).
49. J. Mickelsson, *Comm Math. Phys.*, *110*: 173 (1987).
50. R. V. Moody, *Bull. Am. Math. Soc.*, *73*: 217 (1967).
51. R. V. Moody, *A Simplicity Theorem for Chevalley Groups Defined by Generalized Cartan Matrices*, preprint (1982).
52. D. Munford, *Tata Lectures on Theta*, Birkhaüser, Boston (1983).
53. M. K. Murray, *Comm. Math. Phys*, *116*: 73 (1988).
54. A. Pressley and G. Segal, *Loop Groups*, Oxford University Press, Oxford (1986).
55. A. Rocha-Caridi, MSRI Publ. *3* (1985).
56. A. Rocha-Caridi and N. R. Wallach, *Invent. Math.*, *72*: 57 (1983).
57. G. Segal, *Comm. Math. Phys.*, *80*: 301 (1981).
58. N. N. Shapovalov, *Funct. Anal. Appl.*, *6*: 307 (1972).
59. I. Singer, *Elie Cartan et les Mathématiques d'Aujourd'hui*, Proceedings of the Conference, Lyon (1986).
60. T. H. R. Skyrme, *J. Math. Phys.*, *12*: 1735 (1971).
61. P. Sorba and B. Torresani, *Int. J. Mod. Phys. A*,*3*: 1451 (1988).
62. H. Sugawara, *Phys. Rev.*, *170*: 1659 (1969).
63. J. Thierry-Mieg, *Nonperturbative Quantum Field Theory*, Cargese Lecture Notes (G. t'Hooft, A. Jaffe, G. Mack, P. K. Mitter, and R. Stora, eds.) pp. 567–576 (1988).
64. J. Tits, *Groups and Group Functors Attached to Kac-Moody Data*, Lecture Notes in Math, vol. 1111 Springer-Verlag, Berlin and New York pp. 193–223 (1984).
65. J. Tits, *Groupes Associes aux Algèbres de Kac-Moody*, Seminaire Bourbaki 700, Hermann, Paris (1988).
66. B. Torresani, Ph.D. thesis, CPT-86/P. 1893 (1986).
67. B. Torresani, *Lett. Math. Phys.*, *13*: 7 (1987).
68. A. Tsuchyia and Y. Kanie, *Duke Math. J.*, *53*: 1013 (1986).
69. V. S. Varadarajan, *Lie Groups, Lie Algebras and Their Representations*, Springer-Verlag, Berlin (1984).
70. M. Vergne and H. Rossi, *Acta Math.*, *136*: 1 (1976).
71. N. D. Verma, *Bull. Amer. Math. Soc.*, *74*: 160 (1968).
72. A. M. Vershik, I. M. Gelfand, and M. I. Graev, *Comp. Math.*, *35*: 299 (1977).
73. A. B. Zamolodchikov, *Th. Math. Phys.*, *65*: 1205 (1965).

Index